Introduction to Biology: Laboratory Exercises

Seventh Edition

R. W. Peifer

University of Minnesota
Minneapolis, Minnesota

Cover Photo Acknowledgment
The photograph on the cover of the lab manual was taken by Professor Richard Phillips, Department of Ecology, Evolution, and Behavior, University of Minnesota. This superb photo shows a clump of Showy Lady's-slipper, *Cypripedium reginae,* Minnesota's state flower. The photo was taken near Lake Itasca, the headwaters of the Mississippi River.

Copyright © 1997, 1995, 1990, 1985, 1982, 1979, 1978 by R.W. Peifer

ISBN 0-8087-3440-7

All rights reserved. No part of this book may be reproduced in any form whatsoever, by photography or xerography or by any other means, by broadcast or transmission, by translation into any kind of language, nor by recording electronically or otherwise, without permission in writing from the publisher, except by a reviewer, who may quote brief passages in critical articles or reviews.

Printed in the United States of America.

J I H G F E D C B

Address orders to:

BURGESS INTERNATIONAL GROUP, INC.
7110 Ohms Lane
Edina, Minnesota 55439-2143
Telephone 612/820-4561
Fax 612/831-3167

Burgess Publishing
A Division of BURGESS INTERNATIONAL GROUP, Inc.

Acknowledgments

I would like to express my gratitude to Bruce Fall for his many valuable suggestions and assistance in preparing this edition. Special thanks also to Jessica Kauphusman and Geri Grosinger for help in proofing the manual.

My gratitude to Dr. M. Blumenfeld and Dr. W. Rottman for providing the outline of the experimental protocols used in the enzyme exercise.

"The obscure we eventually see, the completely apparent takes longer."

Anonymous

Contents

1. The Microscope 1
2. Microscopic Organisms 13
3. Diffusion and Osmosis 35
4. Enzymes 47
5. Respiration 67
6. Photosynthesis 93
7. Mitosis and Meiosis 111
8. Genetics 127
9. Plant Biology 159
10. Plant and Animal Adaptations 185
11. Origin of Life 193

Appendix I: Intra-Conversion in the Metric System 207

Appendix II: Theory and Operation of the Spectrophotometer 209

General Laboratory Instructions

1. Coats and miscellaneous items should be placed on the coat rack, not on the lab table top.

2. No smoking or eating in the lab at any time.

3. Check the bulletin board for last minute instructions and special announcements each day when you arrive.

4. You are responsible for the safe operation of your microscope. When you are finished using the scope, please return it to the scope cabinet.

5. When working with flammable solvents, use small amounts and only in the fume hood.

6. Use caution when using boiling water and caustic reagents.

7. When open flames are being used in lab, be careful that loose strands of hair do not fall into the flames.

8. Turn off hot plates when you are through using them.

9. Make certain you leave your work space and any instruments you have used in clean and orderly condition.

    a. Pick up waste paper and place in trash container.

    b. Clean your microscope at the end of the lab period by using lens paper and the appropriate cleaning solvent.

    c. Wash all glassware when you finish with it. Make certain that the sinks do not become clogged with cover slips, waste paper or animal tissues.

1 Microscope

INTRODUCTION

Many aspects of biology deal with structures that are so small that they cannot be visualized with the unaided eye. Many generalizations in biology would never have been formulated had it not been for the development of the **light microscope**. For example, to arrive at the generalization that all organisms are composed of cells, and that cells arise only from preexisting cells required a method of looking at objects of extremely small dimensions. With the development of the light microscope as a research tool in the early seventeenth century, the means were available to formulate this generalization and provide the necessary supporting evidence.

The light microscope has served humankind extremely well for over 300 years, but is limited in its ability to resolve objects that are smaller than 0.25 micrometers (See Appendix I for information on making conversions within the metric system). This limitation in resolving power is due to the long wavelength of visible light used to illuminate specimens. Much of the internal structure of the cell smaller than 0.25 micrometers remained a mystery until the development of the electron microscope in the early 1930's. The electron microscope has extended human perception many orders of magnitude beyond that of the light microscope because it uses a beam of electrons with an extremely short wavelength to illuminate objects. Today there are electron microscopes that can resolve objects that are over a single angstrom in dimension. Though electron microscopes can resolve objects that are very small, they cannot be used to view living specimens because the specimens must be coated with a metallic film. New imaging tools, such as the scanning-tunneling microscope, are creating new perspectives of the microscope world.

The light and electron microscopes are only a small number of the instruments available to scientists today in their search for explanations of biological phenomena. Other techniques and equipment include liquid scintillation counters, ultracentrifuges, isotopic traces, spectrophotometers, and numerous chromatography methods, of which you will use the latter two in future laboratory exercises.

During today's exercise, you will begin your investigations into small dimensions with the aid of the light microscope. Unfortunately, because of sophistication and cost, we cannot provide you with an electron microscope. Even though the light microscope is limited in its resolving power, it remains an important and an impressive research tool. It should prove adequate in helping you develop an appreciation for the dimensions within which science operate. At a minimum, it is a point for departure.

OBJECTIVES

1. Identify the major parts of a compound microscope.
2. Learn how to operate a compound microscope.
3. Measure the diameter of field of three different fields of view, and use these measurements to determine the length and width of objects in millimeters and micrometers.
4. Learn how to convert metric values within the Metric System.
5. Understand **depth of focus** and **diameter of field** and how they are affected by different magnifications.
6. Compute total magnification using different objective lenses.
7. Show competency in the use of the compound microscope by locating and correctly identifying specific plant and animal cell structures.
8. Be able to describe basic plant and animal cell structure.

MATERIALS

compound microscopes
lens cleaning solvent
lens paper
microscopic slides
coverslips
toothpicks
prepared slides:
    printed letters
    colored threads (three colors)
    microscope calibration slide (1 division = 0.10mm)
    methylene blue stain
living specimens:
    Anacharis sp.
    human cheek cells

PROCEDURES

General Instructions

You are responsible for any damage to your microscope during your laboratory period, so it is to your advantage to observe the following directions.

1. When obtaining a microscope from the microscope cabinet, always pick up the scope with two hands, one hand on the arm and the other supporting the base.
2. If lenses (objective or ocular) are dirty, clean them only with lens paper and a cleaning solvent. Do not use substitutes.
3. If an object is in focus while using a lower power objective, the higher power objectives should clear the microscope slide when rotated into viewing alignment. The objectives are said to be **parfocal** when only small adjustments are needed to refocus objects while changing magnifications. Use only the fine adjustment when focusing with higher power objectives.

4. If you are using oil immersion, be certain that you do not rotate the 4X, 10X, and 40X objectives into oil. These objectives are not designed to be immersed in oil, if they are, they can be permanently damaged. Please use caution.
5. Always clean the oil immersion objective with the cleaning solvent provided before returning your scope to the scope cabinet.
6. Never grab the objective lenses to change magnification, only rotate the nosepiece by grasping the ring above the objective lenses.
7. Store your microscope with the scanning objective (4X) rotated into viewing alignment.
8. If your microscope is not functioning properly, do not struggle with it, ask your instructor for assistance.

I. PARTS AND OPERATION OF THE COMPOUND MICROSCOPE, Figure 1-1

Figure 1-1a and 1-1b are diagrams of a compound microscope with some of the major components labeled. Most modern compound microscopes follow this basic design. A microscope is **compound** if it uses two lenses in a series to produce magnification. The lens nearest the eye is called the **ocular** and the other lens the **objective**. Your microscope has three dry objective lenses (4X, 10X, 40X), and one oil immersion lens (100X) that can be rotated into alignment with the ocular lens and barrel. The numerical and letter designations assigned to each objective lens, 4X, 10X, 40X, and 100X mean the lenses can magnify objects 4, 10, 40 and 100 times, respectively. The ocular lens in your microscope can magnify objects 10 times (10X). A compound microscope combines the magnifying capabilities of the ocular and objective lenses. The total magnification of the two lenses together is determined by multiplying their individual magnifying capabilities (e.g., 10X x 4X = 40X). In the preceding example, the two lenses together can magnify objects 40 times.

Calculate the magnifying capabilities of the following:

Ocular Lens		Objective Lens		
10X	x	4X (scanning)	=	_____
10X	x	10X (low power)	=	_____
10X	x	40X (high power)	=	_____
10X	x	100X (oil immersion)	=	_____

The 100X oil immersion objective should only be used with immersion oil. It should be cleaned with the proper solvent before you put your scope away.

CAUTION: This is the only lens that you should use with immersion oil. If you inadvertently get immersion oil on any of the non-oil immersion lenses, ask your instructor for assistance.

Exercise 1—Microscope 3

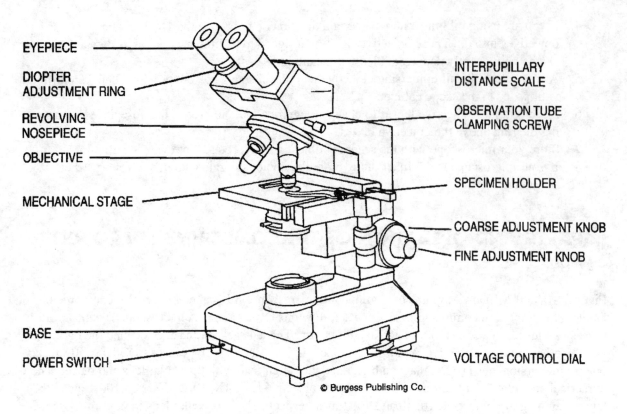

FIGURE 1-1a. Compound Microscope

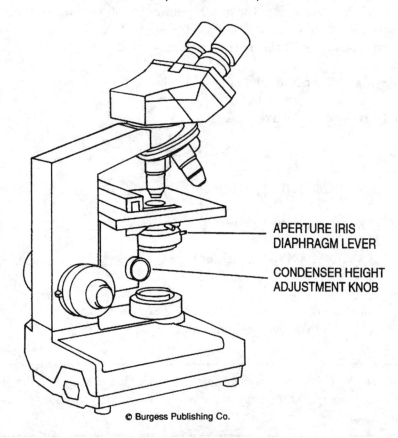

FIGURE 1-1b. Compound Microscope

4 Introduction to Biology: Laboratory Exercises

The ocular lens contains a pointer that rotates. When you have a question concerning an object or structure you are viewing and want your instructor to look at it, place the object at the tip of the pointer.

Light microscopes are limited in their capability to magnify and simultaneously **resolve** objects by the wavelength of radiant energy used to illuminate the specimen. The resolving power of a microscope is its capacity for separating two points that are very close together so they appear as two points rather than one. The resolving power of a microscope increases, as the wavelength of light decreases The best light microscopes that use white light for illumination cannot distinguish two lines as separate lines if the lines are placed closer together than 0.25 μm, even at a magnification of over 1000X. The limit of resolution for your microscope probably is about 2–3 μm.

Because light must pass through specimens, only objects that are translucent can be viewed with the light microscope. Therefore, specimens must generally be extremely thin and stained with some type of dye.

Figure 1-2 shows how light passes through a compound microscope. Light from the illuminator passes, successively, through the iris diaphragm, condenser, microscope slide and specimen, objective lens, and ocular lenses to the eye. The **iris diaphragm** regulates the amount of light that reaches the specimen and should be continually adjusted to give the optimal amount of light. The **condenser** focuses the light through the specimen. For most viewing it should be rotated upwards until it stops or makes contact with the bottom side of the microscope slide. This will place the condenser just under the microscope slide. In some situations, the condenser will need to be rotated slightly downward. The **coarse** and **fine adjustment knobs** are used to focus the image of the specimen on the retina of the eye. When the image of a specimen reaches the eye it is reversed (right is now left and top is now bottom). Not only is the image reversed, but so is its movement.

Place a microscope slide with printed lettering on the microscope stage. The coverslip (coverglass) should always be on the upper surface of the slide. Turn the light source on and turn it to approximately 3/4 power. Turning the iris diaphragm lever counterclockwise reduces the diaphragm opening. Adjust the diaphragm so it is just slightly open. Using the mechanical stage to move the slide, center one of the letters on the slide so that the beam of light coming through the condenser passes through the specimen. Rotate the scanning objective (4X) into place. Then, using the coarse adjustment knob, rotate the microscope stage upward until it stops. Note which way to turn the adjustment knob to lower or raise the stage. Look through the ocular lenses. A circular field of light should be visible. If not, make certain that the objective lens has snapped into alignment. While looking through the scope, adjust the distance between the ocular lenses so it matches your interpupillary distance (the distance between your pupils). To make this adjustment so you have correct binocular vision, slide the ocular lenses inward or outward until you see a single circular field of light. Note the number on the scale between the ocular lenses. Whenever you use the scope again, you should automatically set the interpupillary distance to this setting.

Look through the microscope and slowly lower the stage with the coarse adjustment until the lettering is roughly in focus. Then focus with the fine adjustment knob to bring the letter into sharp focus. You may need to make a diopter adjustment to obtain the sharpest focus This adjustment compensates for differences in your eyes. While looking through the microscope, close your left eye and focus the scope with the fine adjustment to get the sharpest image. Without changing the focus, open your left eye and close your right eye. Adjust the diopter focusing ring on the left ocular lens to obtain the sharpest focus possible. You should now be able to view the specimen in sharp focus with both eyes simultaneously. You will need to make this adjustment each time you use the microscope.

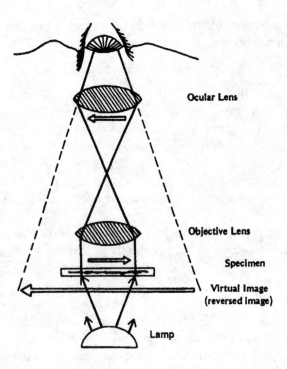

FIGURE 1-2. Path of Light through a Compound Microscope

Notice that the image of the letters is reversed and inverted. Slowly move the slide away from you while looking into the microscope. Which way does the image move? Move the slide to the left. Which way does the image move? Practice focusing your microscope at different magnifications with this slide, while simultaneously adjusting the diaphragm and the voltage control dial to obtain the best lighting effect.

II. DEPTH OF FIELD (FOCUS), Figure 1-3

The center of focus for each magnification remains a fixed point while you are raising and lowering the microscope stage. For an object to be in focus, the specimen must be within the area immediately above or below the center of focus, this is called the **depth of field**. The depth of field in which objects are in focus decreases as magnification increases (Figure 1-3). Place a slide of colored threads on the stage. Under all magnifications, focus up and down with the fine adjustment and try to determine, using depth of field, which thread is on top, which is in the middle, and which is on the bottom. Hint: If the stage is rotated to the maximum upward position and then slowly rotated down, which thread (top, middle, or bottom) would be the first to come into focus? Which magnification works best to determine the order of threads?

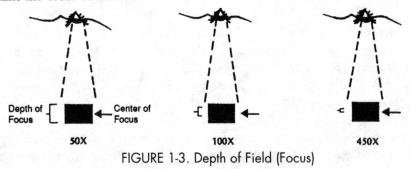

FIGURE 1-3. Depth of Field (Focus)

III. CALIBRATION OF THE MICROSCOPE FIELD, Figure 1-4

In this exercise you will measure the distance across the field of view at each of the following magnifications (40X, 100X, and 400X). You will not measure the diameter of field under immersion oil. Once you have determined the diameter of field at each magnification, you will be able to use this information to estimate the dimensions of organisms and structures viewed through the microscope.

It should be apparent by now that as you increase magnification the amount of area you can observe decreases. In other words, as magnification **increases** the diameter of field **decreases**.

Place a calibration slide on the microscope stage, and rotate the 4X objective into place. You can think of this calibration slide as a miniature ruler that you will use to measure the diameter of field at each magnification. The length of the ruler is 1.0 cm, with the smallest divisions on the ruler equal to 0.10 mm. NOTE: A division is the distance between the midpoints of two adjacent parallel lines or the distance between the left edges of two adjacent parallel lines.

To measure the diameter of field at 40X, align the left edge of the black line marked zero so that it forms a vertical tangent to the left outer edge of the microscope field. You should now see an image similar to Figure 1-4. Count the number of divisions across the diameter of the field. Record this value in Table 1-1. Now, without moving the slide, swing the 10X objective into viewing alignment. Again position the line marked zero tangent to the left outer edge of the microscope field. Notice that you are now viewing only a portion of division that you counted at 40X. Count the number of divisions at this magnification and enter the value in Table 1-1. Repeat this procedure at 400X.

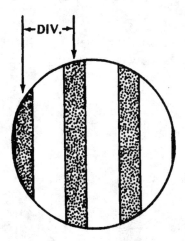

FIGURE 1-4. Example of Divisions per Field

To convert the number of divisions per field to the number of millimeters per field, simply multiply the number of divisions times the number of millimeters per division, in this case 0.10 mm. You can use the following formula.

$$\text{millimeters per field} = \text{\# divisions per field} \times \text{millimeters per division (0.10 mm)}$$

One millimeter equals 1000 micrometers, so to convert the number of mm/field to the number of μm/field you multiply the mm/field by 1000 (see Appendix I, **Intraconversion in the Metric System**).

What would you have to do to increase the accuracy of your estimates at a particular magnification?

What is the difference between a precise measurement and an accurate one?

Table 1-1. Measurement of Diameter of Field				
Ruler—mm/div.	**magnification**	**div./field**	**mm/field**	**µm/ field**
0.10 mm	40X			
0.10 mm	100X			
0.10 mm	400X			

Because of the inversely proportional relationship between magnification and diameter of field, you can mathematically calculate the diameter of field of any magnification when you know the diameter of field of a single magnification. You can use the following formula to calculate the diameter of field for other magnifications.

$$\text{unknown diameter of field} = \text{known diameter of field} \times \frac{\text{magnification of known diameter of field}}{\text{magnification of unknown diameter of field}}$$

For example, if you have directly measured the diameter of field at 40X and found it to be approximately 4500 micrometers, you can then simply use the formula above to determine the diameter of field at a magnification of 400X or any other magnification:

Substituting you would have:

$$\text{unknown diameter of field} = 4500\mu m \times \frac{40X}{400X}$$

Solving you would have:

$$\text{unknown diameter of field} = 4500\mu m \times \frac{1}{10} = 450\mu m$$

Note that in this case, when the magnification was increased 10 times, the diameter of field decreased by the same factor, that is, by a factor of 10.

Using the same approach, calculate the diameter of field at 100X magnification. What is the diameter of field at 100X? _____ µm; _____ mm. How well do these values correspond to the values you obtained through direct measurement? Can you account for the differences in these values?

You will want to remember the approximate values for the three magnifications so that you can determine the dimension of objects viewed in future exercises. You can think of the diameter of field you calculated or measured for each magnification as an invisible ruler superimposed across the respective diameter of field. Try your skill at determining the size of an object. Obtain a prepared slide that has printed lettering on it. At 40X, estimate the height and width of the first capital letter.

Express your results in both mm and µm.

    Height _____ mm, _____ µm
    Width _____ mm, _____ µm

IV. DEVELOPING SKILLS WITH A COMPOUND MICROSCOPE, Figures 1-5, 1-6

You will now look at representative types of cells from two of the kingdoms, Plantae and Animalia, to help you further develop your proficiency in using a microscope. Organisms from these kingdoms are eukaryotic, so the cells have a number of basic structural features in common.

The outer most living boundary of all cells, even prokaryotes, is called the **plasma** or **cell membrane**. Material can be deposited to the outside surface of the plasma membrane by the cell, but this material is not considered to be part of the living cell (**protoplasm**). All eukaryotic cells have the bulk of their hereditary material (DNA) enclosed within a **nuclear membrane**. This membrane is structurally similar to the plasma membrane, but is two layers thick (double membrane) and forms the outer boundary of the structure called the **nucleus**. The most prominent feature within the nucleus is the **nucleolus** (plural; nucleoli), the site of ribosomal RNA synthesis. The dark staining granular material is condensed genetic material called **chromatin**.

Outside the nucleus is the **cytoplasm**. This is the portion of the cell that includes everything outside the nuclear membrane including the plasma membrane. Found in the cytoplasm are the other membrane-bound organelles that characterize eukaryotic cells. These include: **mitochondria, Golgi apparatuses, rough and smooth endoplasmic reticulum, vesicles, microfilaments, microtubules, and vacuoles**. Most of these structures are too small to be seen with a light microscope.

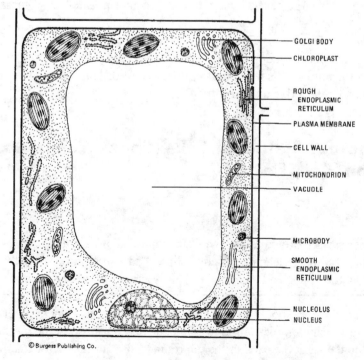

FIGURE 1-5. Plant Cell

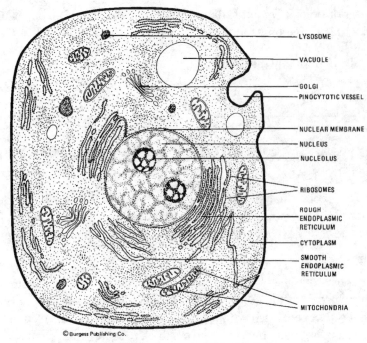

FIGURE 1-6. Animal Cell

Plant and animal cells have all of the above features in common, but differ in the following ways. Plant cells contain membrane-bound organelles called **plastids** (Figure 1-5). The most common type of plastid is called the **chloroplast**. Chloroplasts contain the photosynthetic pigments that are used by plants to capture light energy. Because plants contain these organelles, they are known as **photoautotrophs**. Photoautotrophs are able to manufacture food molecules using inorganic materials and light energy. Plant cells also deposit extracellular material outside the plasma membrane that forms a **cell wall** that provides support and protection. The cell wall is composed largely of cellulose. Another prominent feature is a large **central vacuole** that occupies most of the volume of a typical plant cell.

Animal cells lack cell walls, plastids, and large central vacuoles (Figure 1-6). There is extracellular material outside the plasma membrane called the **glycocalyx**, but it is structurally unlike the cell walls of plants. Because animal cells lack photosynthetic plastids, animal cells must feed upon other organisms, and are referred to as **heterotrophs**.

A. Squamous Epithelial Cells (Cheek Cells)—Example of an Animal Cell, Figure 1-7

Procedure

Scrape the inner surface of your cheek with the blunt end of a toothpick and spread the scrapings into a small drop of water on a clean microscope slide. Add a small drop of methylene blue solution to the cells, mix with a toothpick, add a coverglass, and examine under the microscope. The methylene blue

 CAUTION: You should handle and dispose of your toothpicks safely. Do not share toothpicks or touch toothpicks used by someone else. Dispose of the toothpicks in the plastic lined container at the back of the lab room. Ask your instructor for specific directions.

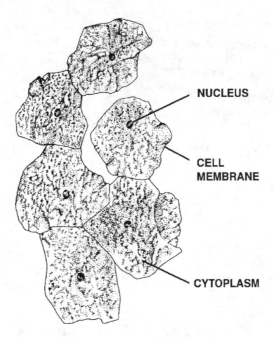

FIGURE 1-7. Human(Epithelial) Cells

dye preferentially stains the nuclei of cells. Draw a stained cell and label the structures you can identify. What is the diameter of your cheek cells?

_____mm, _____μm.

What structures are present or absent in this cell that allow you to distinguish it as an animal cell?

B. *Anacharis* Leaf Cells (Water Weed)—Example of a Plant Cell, Figure 1-8

Procedure

Select a bright green *Anacharis* leaf from near the tip of a sprig and mount in a small drop of water, cover with a coverglass and examine under all magnifications. Identify the cell walls, vacuoles, nuclei, and chloroplasts of the cells. How many cell layers thick is the leaf?

> ☞ **HINT**: Use the depth of field at 400X to determine this.

The leaf is colored by chlorophylls, green photosynthetic pigments, which are localized within the chloroplasts. At 400X, note the movement of the chloroplasts within the cells. This movement is

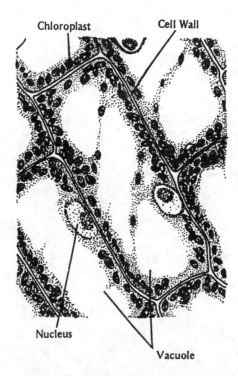

FIGURE 1-8. Anacharis Leaf Cells

Exercise 1—Microscope 11

called **cytoplasmic streaming** and occurs widely in plant and animal cells. Speculate as to why this streaming occurs. (HINT: Is there some advantage to the cell?) Draw and label a single cell of the leaf. What is the approximate length in micrometers?
_____μm

What is the approximate width in micrometers? _____μm

Indicate with arrows the direction of cytoplasmic streaming. What structures are present or absent in this cell that allow you to distinguish it as a plant cell?

2 Microscopic Organisms

INTRODUCTION

Approximately 1.5 million organisms have been described by biologists. Many believe that there are millions of species yet to be discovered, most living in the tropical rain forests of the world. Some estimates of the total number of extant species run as high as 40 million, with insects being the vast majority. It should be obvious that in trying to describe millions of organisms, a systematic method of naming and arranging organisms is essential. **Classification** is the process by which organisms are sorted and placed into groups (taxa). The underlying rationale for creating these groups comes from the area of biology called **systematics**. The categories of organisms that emerge from a contemporary classification system hopefully reflect more than casual or artificial relationships within and among groups. Classification schemes are assembled using the most current information that reflects the evolutionary relationships of organisms. Such an orderly system is possible because the evolutionary process has been simultaneously responsible for the uniformity and diversity of the living world.

It is probably a safe assumption that most people are much more aware of the differences (diversity) among organisms than they are aware of their similarities (uniformity). Some similarities among organisms are readily apparent, while others are not so obvious. For example, we categorize all organisms that possess feathers as birds, but it might not be so obvious why whales and humans would also be placed in a common category. On close examination we find that whales and humans possess hair and mammary glands, and are therefore placed in an inclusive group called the mammals. Though the gross shape and form (morphology) of a whale hardly resemble what most people think a typical mammal should look like, it is nevertheless established that a whale and a man share more in common than a whale and a shark. In the same vein, we could easily place bats, birds and butterflies, all of which have wings and fly, in a common group for classification purposes. This system would not be very revealing nor satisfying in its attempt to explain the differences and similarities of these organisms. Therefore, we see that the presently accepted scheme of classification is based not on single artificial criteria, but on a multitude of characteristics that reflect evolutionary relationships and history.

Much of the success in explaining the evolutionary relationships among organisms has been due to the refinement of techniques in chemistry that can be used to identify the structure and function of substances found in organisms. Is it any wonder that modern biology takes such a biochemical approach in many of its investigations?

Organisms must carry on a fantastic array of processes to survive, and much of what they possess in common is reflected in these processes. Some of these processes include procurement of raw materials, chemical conversion of raw materials into energy and utilizable compounds, voiding of nonuseable materials and metabolic wastes, exchange of gases, transport of materials within the organism, sensing the environment (both external and internal), regulating water balance, reproduction, and many more. A universal feature of organisms is that they are composed of **cells** (see Cell Theory in your text). The cell is the fundamental unit into which organic molecules mold themselves to form a functioning entity that can gather energy from the environment for its maintenance and perpetuation.

There is a great deal of similarity in the components and structure of cells (organelles, biochemical pathways, etc.) at the subcellular level. In the exercises that follow, you will be given the opportunity to briefly survey some microscopic representatives from four of the five major groups into which organisms are classified. Many subcellular structures should be obvious while others may not be so apparent. Much of what you will see depends on how well you have mastered the use of the microscope. Some similarities that exist between cells are not discernible with your microscope (e.g., endoplasmic reticulum, plasma and nuclear membranes, and ribosomes), but their existence has been established with the aid of more sophisticated equipment such as the electron microscope.

Before you begin the following exercises, it might be helpful if you are given an example of how biologists use a classification system. Remember that classification is the arrangement of organisms into categories based on similarities and dissimilarities, and that modern classification systems strive to reflect historical evolutionary relationships. The largest or most inclusive category to which we assign organisms is the **kingdom**. In the mid 1800's, at the time Charles Darwin was making his famous voyage on the H.M.S. Beagle, natural historians (i.e., biologists) classified all life into two kingdoms, plant and animal. Currently most biologists favor a **five-kingdom classification** system as proposed by Robert Whittaker in 1969 (Table 2-1), although many scientists disagree about where major groups of organisms should reside. For example, Whittaker placed all unicellular eukaryotes in their own kingdom Protista. More recently, the trend has been to expand the borders of that kingdom

Table 2-1. Five Kingdom Classification System

Monera	**Protista**	**Plantae**	**Fungi**	**Animalia**
Archaebacteria	Protozoa	Mosses	Zygote fungi	Sponges
Eubacteria	Algae	Liverworts	Sac fungi	Comb jellies
	Funguslike Protists	Hornworts	Mushrooms	Jellyfishes, corals, anemones
		Whiskferns		Flatworms
		Club mosses		Rotifers
		Horsetails		Roundworms
		Ferns		Ribbon worms
		Conifers		Clams, snails, octopuses
		Cycads		Segmented worms
		Ginkgo		Crustaceans, insects, spiders
		Gnetae		Bryozoans
		Flowering plants		Brachiopods
				Phoronids
				Sea stars, sea urchins
				Tunicates, lancelets, vertebrates

to include multicellular organisms. Some of these organisms were previously classified in the kingdoms Plantae and Fungi. As an example, the multicellular algae, including kelp and other seaweeds, are now classified in the kingdom Protista. An additional realignment places some of the fungus-like organisms (plasmodial and cellular slime molds, and water molds) with the Protista.

Taxonomy is like any area of science. As new information is uncovered and discoveries are made, it may become essential to rethink old ideas and form new hypotheses. For example, new information coming from the comparison of DNA and protein sequence analysis is reshaping systematists' view of the five-kingdom classification system. There have been proposals to create a sixth kingdom by splitting the kingdon Monera into two kingdoms, Eubacteria and Archaebacteria. Another proposal would create an eight-kingdom system by splitting the Monera as just described, and split the Protista into three kingdoms, the Archezoa, Chromista, and Protista. Therefore, when biologists find it necessary to realign the boundaries of classification systems because of new evidence, it should not be viewed as confusion, but as an integral part of the process of science.

Each category in a classification system is arranged in a hierarchy with each group more inclusive (having a greater variety of characteristics) than the group directly beneath it. For example, humans are classified as follows:

Kingdom	Animalia
Phylum	Chordata
Class	Mammalia
Order	Primates
Family	Hominidae
Genus	*Homo*
Species	*sapiens*

Under this system of classification we belong to the class Mammalia that contains all animals that possess mammary glands and hair. The next category, order, places us in a more exclusive group—the Primates (Old and New World monkeys, apes, lemurs, marmosets, etc.). The two part name (**binomial**) which describes our species is *Homo sapiens*. The first name is the **genus** (plural, **genera**) into which species are placed. The second word is the **specific epithet**. Together the genus and specific epithet create a unique scientific name for the species. It is convention the first letter of the genus be capitalized, and the entire scientific name be either underlined or italicized.

More complex classification systems exist, but all follow the basic design given above. In these comprehensive classification arrangements, additional categories are created by adding the prefixes super-, sub-, and infra- to the main groups. For humans, the classification for the category class would look like this:

Superclass	Tetrapoda
Class	Mammalia
Subclass	Theria
Infraclass	Eutheria

Sometimes problems arise in classification when an organism possesses major characteristics that are criteria for assignment to two large categories. *Euglena* exemplifies this problem. When you first observe *Euglena*, their ability to swim using a whip-like structure called a flagellum will probably intuitively tell you that they are "animal-like." On closer examination you would also find they possess chlorophyll in structures called chloroplasts, and that they can manufacture their protoplasm from inorganic sources—a truly "plant-like" activity. Then again, they lack cell walls, a distinguishing characteristic of true plants. Where would you place *Euglena* in a classification system?

In the following exercises, you will look at microscopic representatives from four (Monera, Protista, Fungi, and Animalia) of the five kingdoms.

OBJECTIVES

1. Survey microscopic representatives from four of the five kingdoms.
2. Identify and know the function of certain cell organelles.
3. Estimate the dimensions of the organisms you observe.
4. Define and apply the following terms:

procaryotic	pseudopodia
eukaryotic	contractile vacuole
autotrophic	nucleus
heterotrophic	chloroplast
cell theory	nucleolus
classification	macronucleus
phagocytosis	trichocysts
pinocytosis	plasma membrane
flagella	cilia

5. Compare the means of locomotion and energy procurement processes of the organisms you observe.
6. Explain how certain cell structures are related to function.

MATERIALS

living specimens:
    Gloeocapsa sp.   *Amoeba sp.*   *Aeolosoma sp.*
    Paramecium sp.   *Blepharisma sp.*   *Volvox sp.*
    Spirogyra sp.   *Rhizopus sp.*   *Euglena sp.*
    Cyclops sp.   *Daphnia sp.*
    mixed culture of green algae
    mixed culture of diatoms
    pond water
    undyed yeast

prepared slides of bacteria   prepared slides of diatoms
microscope slides   coverslips
compound microscopes   microscope depression slides
methyl green stain   methylene blue stain
methyl cellulose   toothpicks
red dyed yeast   lens paper
lens cleaning solvent   immersion oil
dissecting microscopes

I. MONERA—Eubacteria

The first major division and largest grouping you can make among all organisms is based on the presence of a membrane-bound nucleus (**eukaryotic**) or the lack of it (**prokaryotic**). The kingdom monera contains only prokaryotic organisms (generally called bacteria). Besides lacking a nuclear membrane, prokaryotes also lack mitochondria, chloroplasts, and other membrane-bound organelles that are characteristic of eukaryotes.

Historically, the classification of the monera has been largely artificial. Biologists used characteristics, such as the size and shape (morphology) of cells, mode of motility, mode of nutrition, and the presence or absence of spores as classifying criteria. None of these criteria necessarily reflected evolutionary connections. Therefore, the classification system created was not hierarchical nor based upon evolutionary relationships. Nonetheless, this artificial system has been extremely useful in medicine, agriculture, and industry.

With recent advances in the study of intracellular structure and biochemistry, new information is emerging that allows biologists to reconstruct the evolutionary relationships (**phylogeny**) among the Monera. One discovery is that a few groups of prokaryotes are distinctive from the rest. These groups include the thermoacidophiles (bacteria that live in environments that are highly acidic and hot), methanogens (bacteria that make their living making methane from carbon dioxide and hydrogen gas), and halobacteria (bacterial that live in extremely salty habitats). These groups now make up a major subdivision of the Monera called the Archaebacteria. The remaining prokaryotes are placed in a group called the Eubacteria ("true bacteria").

The Monera are the most numerous organisms in the world. A typical gram (1/28 oz.) of fertile soil can contain billions of individual bacterial cells. This group contains many ecologically diverse types, from those living in hot sulfur springs, to those that make their living at the depths of the oceans where light does not penetrate. Some cause disease (e.g., *Streptococcus pneumoniae*) like pneumonia, while others like the nitrogen fixers of the genus *Rhizobium* are beneficial to humans and other organisms.

The group is also metabolically very diverse, containing individuals that obtain their energy by oxidizing inorganic substances like ammonia, nitrates, sulfides, or iron. These types of organisms are called **chemoautotrophs**. Probably the largest group of bacteria are the **heterotrophs**. Heterotrophs obtain their energy by consuming food molecules produced by other organisms. Because these organisms have no mechanism for ingesting large food particles, they must absorb small organic molecules from their environment. Organisms that absorb food from their surroundings are called **saprobes**. Many members of this group of bacteria are called the **decomposers** because they convert the carbon of dead organisms to carbon dioxide, and release the other elements that make up organisms to the environment. They are the ultimate recyclers.

The Monera also contain groups that are **photosynthetic autotrophs**, organisms that can make organic compounds from sunlight, carbon dioxide, and an electron donor, such as water or hydrogen sulfide. The Cyanobacteria (formerly blue-green bacteria), which you will look at in this exercise, belong to this group.

Bacteria are living everywhere in our environment, and they seem to have a diverse appetite for some most unlikely of foods. We have discovered bacteria that can "eat" oil, pesticides, herbicides, and even toxic substances like polychlorinated biphenyls (PCBs). The new techniques of recombinant

DNA are allowing scientists to transfer the genetic material responsible for these abilities to different microbes, creating "new" bacteria that can be used to attack certain environmental problems, such as oil spills.

A. Heterotrophic Bacterial Types, Figure 2-1

As mentioned earlier, many bacteria that cause disease or have agricultural or industrial importance have been classified based upon morphology. The following are examples of some common heterotrophic bacteria. There are the rod-shaped bacteria (bacilli), such as *Escherichia coli* that live in your intestinal tract, the spherical forms (cocci), such as *Staphylococcus epidermidis* that live on your skin, and the spiral forms (spirilla), such as *Treponema pallidum* that causes syphilis.

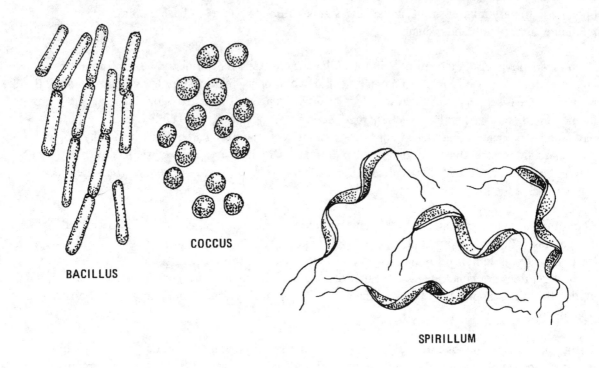

FIGURE 2-1. Bacilli, Cocci, and Spirilla Bacteria

Procedure

Obtain a prepared slide of bacteria. The slide contains all three types of bacteria (bacilli, cocci, and spirilla). Look at the specimens under 40X, 100X, and 400X. When you have viewed all three types of bacteria under these magnifications, look at the microbes under 1000X. The 100X objective requires the use of immersion oil. Before you use the oil immersion objective, you must first finely focus on one type of bacteria using the 40X objective. Without changing the focus, swing the 40X objective halfway out of alignment, but do not swing the 100X objective into place. Looking at the microscope stage from the side, place a small drop of immersion oil on top of the coverglass over the specimen area. Slowly rotate the 100X objective into alignment until you hear it click into place. Because the lenses are parfocal, you can probably find your specimen by making only small changes with the fine focus knob. You may need to increase the light intensity and increase the opening of the iris diaphragm. Look at all three types using 1000X.

Notice that even at 1000X the internal cell looks amorphous, this is because prokaryotes lack membrane-bound organelles and internal compartmentalization.

B. Photoautotrophic Bacteria (Cyanobacteria)— *Gloeocapsa*, Figure 2-2

The cyanobacteria include those bacteria that can make their own food from inorganic compounds, needing only the raw materials of carbon dioxide, minerals, water, and light energy. Organisms that can manufacture their food from inorganic sources are called **autotrophs**. Because the process in the cyanobacteria is driven by sunlight, the cyanobacteria are called **photoautotrophs**. The cyanobacteria possess the green photosynthetic pigment chlorophyll *a* and the blue pigment phycocyanin, therefore their name, blue-green bacteria. They were formerly called blue-green algae, but we now know that they have greater affinity with the monera.

Procedure

Prepare a wet mount of *Gloeocapsa* and observe it at all magnifications. *Gloeocapsa* is frequently found on moist rocks and walls, on flower pots in greenhouses, and on wet lake docks. In *Gloeocapsa*, the process of photosynthesis (capturing radiant energy and converting it to chemical energy) occurs only in the cytoplasmic matrix.

The gelatinous sheath is extracellular material secreted by the cells. It is sometimes difficult to find a single cell because the daughter cells cohere temporarily at the end of cell division. When groups of cells are together, it may appear they are surrounded by a single, common sheath, but on closer examination it will be revealed each cell has an individual sheath. Common sheaths are of parental origin.

Notice the cytoplasm of *Gloeocapsa* appears amorphous (i.e., lacking organization). This is characteristic of most prokaryotic cells. Can you identify any subcellular organelles? Suggest a possible function of the gelatinous sheath.

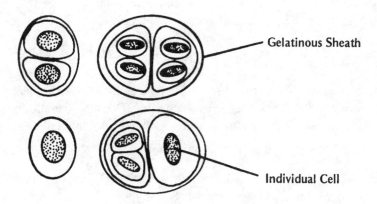

FIGURE 2-2. *Gloeocapsa* (Blue-Green Bacteria)

II. PROTISTA—Algae and Protozoa

The Protista are an ecologically diverse group; wherever you find water, you'll find members of this group. Some live in fresh water while others inhabit marine environments. They can be found in moist soil, and in the rich organic matter of plant communities. Certain members of this group are parasites and cause disease, while others live in more mutual relationships with their hosts.

The group is nutritionally diverse. Some groups are photoautotrophs, while other groups make their living ingesting or absorbing their food. There are even species that combine both methods of nutrition. Most protists are unicellular, but there are numerous species which are filamentous or colonial. There are even some multicellular forms, such as the seaweeds (red and brown algae).

A. Photosynthetic Protists—Euglenoids, Diatoms, Green Algae

The algae are important members of the **phytoplankton**—the photosynthetic organisms that float near the surface of lakes, oceans, and ponds. The phytoplankton form the foundation of most aquatic food chains, and are a major contributor to the oxygen of our atmosphere. Many phytoplankton are kept near the surface of the water where they can capture sunlight for photosynthesis by vacuoles that contain oil. These oil vacuoles make the cells buoyant, and are thought to have contributed to the formation of fossil fuels over geologic time. The group contains the dinoflagellates, most of which are marine. These organisms have cellulose cell walls, two flagella, and the photosynthetic pigments of chlorophyll *a* and *c*. The Euglenoids have no cell wall, 1-3 flagella, chlorophylls *a* and *b*, and most live in freshwater. The main distinguishing characteristic of diatoms is a cell wall composed of mostly silica. They lack flagella, and possess chlorophylls *a* and *c*. There are both freshwater and marine diatoms.

1. *Euglena*, Figure 2-3

Euglena is classified as a unicellular alga in the phylum Euglenophyta. Most members of this group live in fresh water. *Euglena* can acquire energy in two ways. As an **autotroph**, *Euglena* can capture radiant energy and transform it into chemical energy through the process of photosynthesis. If light is absent, some species of *Euglena* can absorb dissolved nutrients or ingest small particles by phagocytosis, thereby satisfying their energy requirements **heterotrophically**.

Procedure

Prepare a regular wet mount of *Euglena*. Observe *Euglena* at all magnifications except at 1000X, and note its general behavior and as many structures as possible. Does the flagellum pull or push *Euglena* through its liquid environment? Indicate the direction of movement by placing an arrow next to Figure 2-3.

Slow the movement of *Euglena* by adding a small drop of methyl cellulose and a small drop of *Euglena* to a fresh slide. Mix and add a coverglass. See if you can locate the structures discussed in the following paragraphs (use 400X).

The anterior end of the organism is the end with the long flagellum. At the base of the flagellum is a pigmented area called the **stigma** or **eye spot** (the spot may appear reddish). The function of the eye spot is not entirely known, but it has been correlated with the **phototactic** response to light (i.e., the movement toward or away from a light source).

Euglena has two flagella, one that is long and emergent, and another that is short and nonemergent. You will not see the shorter flagellum because it is housed in a reservoir at the base of the long flagellum. Posterior to the reservoir area is a **contractile vacuole**. Water from the vacuole is discharged into the reservoir.

There is a single central nucleus that in some species is bounded on either side by large granules of a polysaccharide called **paramylon**. There are usually many chloroplasts of various sizes.

Cell walls are a distinguishing characteristic of plant cells, but they are absent in *Euglena*. An elastic area consisting of the plasma membrane and adjacent flexible protein plates forms the **pellicle**. The pellicle is largely proteinaceous and contains no cellulose. This allows the organism to undergo significant changes in body form called **euglenoid movement**. When the methyl cellulose has retarded the movement of your *Euglena*, estimate the length and width of an individual cell. Express these values in micrometers (do not include the flagellum in your measurements).

Length: _____ μm   Width: _____ μm

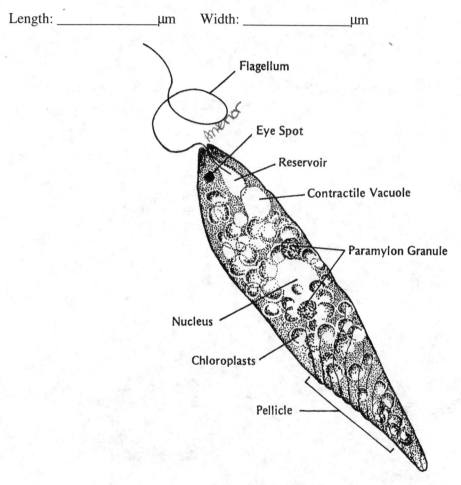

FIGURE 2-3. *Euglena*

2. Diatoms, Figure 2-4

Diatoms are classified in the phylum Bacillariophyta. They exist as single cells, simple chains or loose colonies. They have a cell wall composed mostly of silica (SiO_2), the major component of glass. The cell walls have many elaborate ridges and pits used to identify species. These glass cell walls are very strong and resistant to breakdown. Over geologic time, enormous layers of diatom cell walls have accumulated on the bottom of oceans. Some of these areas have been raised to the surface through geological activity, and are now mined as sources of "diatomaceous earth" for fine abrasives in polishes and toothpastes. Diatoms possess the pigments chlorophyll *a* and *c*, also carotenoids and xanthophylls. They can be found in both fresh water and marine environments.

Procedure

Both prepared slides and fresh cultures of diatoms are available for your observation. Prepare a regular wet mount using the fresh culture. Observe under all dry magnifications (40X, 100X, and 400X). Certain diatoms are capable of movement, and you may see some glide across the slide. The prepared slides contain only the cell walls of the diatoms, the protoplasm is absent or has been shrunken by dehydrating solvents used in the preparation of the slides.

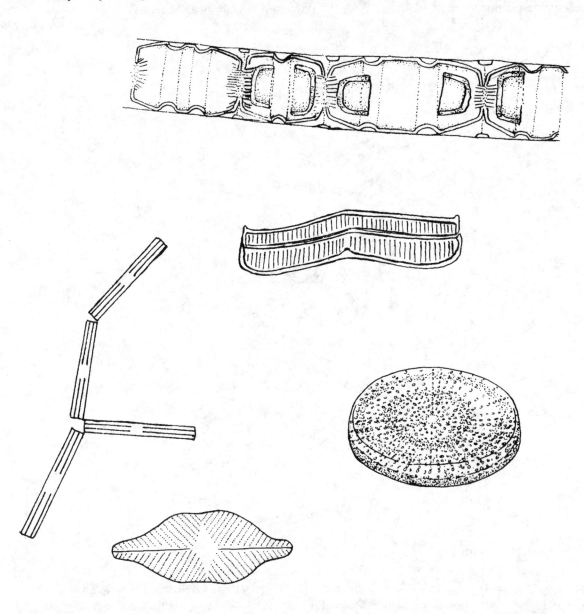

FIGURE 2-4. Various Types of Diatoms

3. *Spirogyra* —A Filamentous Green Alga, Figure 2-5

Spirogyra is a filamentous green alga that contains a spirally arranged chloroplast. It is a freshwater alga common to lakes and ponds. It is a member of the phylum Chlorophyta. Members of this group possess the photosynthetic pigments chlorophyll *a* and *b*, and carotenoids. In addition, members of the Chlorophyta have cellulose as the main component of their cell walls. It is thought that the ancestors of the plant kingdom were members of this group.

Procedure

Prepare a wet mount of *Spirogyra* and observe under all dry magnifications. Search the numerous cells of a filament for a cell that shows most of the cellular structures. You can probably see the spirally arranged chloroplast, the centrally located nucleus, the cytoplasmic strands that support the nucleus, and the large colorless central vacuole. *Spirogyra* provides a good opportunity to observe in detail the basic structure of a plant cell.

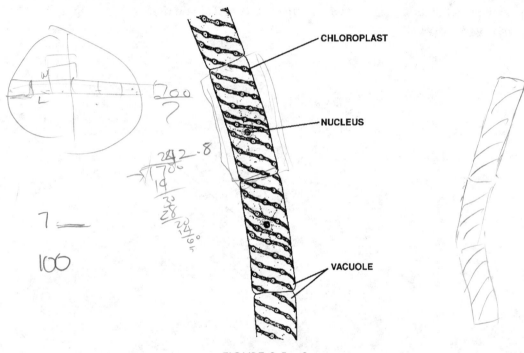

FIGURE 2-5. *Spirogyra*.

Draw and label a single cell of *Spirogyra*. What is the length and width, in micrometers, of a typical cell?

Length: 242.8 μm; Width: 88 μm

Are all the parts of the cell in focus simultaneously at 400X? Why?

4. *Volvox*—A Colonial Green Alga, Figure 2-6

Volvox is a freshwater alga belonging also to the phylum Chlorophyta. It is a colonial alga that may contain many thousands of individual cells embedded in a spherical gelatinous sheath. Each cell has a cellulose cell wall through which a pair of flagella project, and a light sensitive area called a stigma. The cells are connected to adjacent cells by protoplasmic strands. A colony is motile by the coordinated action of the flagella that causes the colony to roll. The colonies can grow to be as large as 0.5 to 1.0 mm in diameter. Some cells within a colony are specialized for specific functions, such as reproduction and photosynthetic activity. Biologists think the simple type of cellular specialization displayed by *Volvox* typifies the early stages toward the evolution of multicellularity.

Procedure

Place a drop of *Volvox* into a depression slide. Fill the entire depression with water so the underside of the coverglass contacts the water when you place a coverglass on top. Begin to view the specimens under 40X. If the colonies are rotating too fast to obtain a good look, remove the coverglass and add a small drop of methyl cellulose. Stir with a toothpick and replace the coverglass. Try to view individual cells at higher magnifications. Remember, your depth of field will be extremely limited at higher magnifications, so you cannot see entire colonies, only individual cells or parts of cells. Speculate as to the function of the light sensitive stigmata. Hint: Do you think it has anything to do with photosynthetic activity?

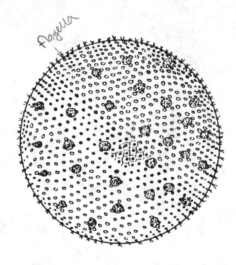

FIGURE 2-6. *Volvox*

5. Mixed Culture of Green Algae

If time permits, you may want to look at other types of green algae. Prepare a regular wet mount from the mixed culture dish, and view under the dry magnifications, especially 40X and 100X. There are freshwater algae guides available to help you identify specific specimens.

B. Heterotrophic Protists—Rhizopoda and Ciliophora

This group contains unicellular organisms called Protozoans. Its members are unable to synthesize organic molecules from inorganic raw materials and an external energy source. They must obtain their food as ready-made sources of energy as organic molecules. Organisms that have this mode of nutrition are called **heterotrophs**. The main method of getting their food is through ingestion.

Six phyla are currently recognized:

Phylum	Means of locomotion	Representative of category
Rhizopoda	Pseudopodia	*Amoeba*
Actinopoda	Floating with pseudopodia	*Actinophrys*
Foraminifera	Pseudopodia	*Globigerina*
Ciliophora	Cilia	*Paramecium, Blepharisma*
Zoomastigophora	Flagella	*Peranema*
Apicomplexa	None	*Plasmodium* (causative agent of malaria—all members of this category are parasitic)

The three means of locomotion for protists are:

Pseudopodia (false feet)—cytoplasmic extensions that function for locomotion and in food procurement.
Cilia—short hairlike structures on the surface of cells used for locomotion and generating water currents that bring food particles in contact with the surface of cells.
Flagellum—a long whiplike extension of the cytoplasm that produces cell movement.

You will look at representatives of two of these groups, the Rhizopoda and Ciliophora.

1. *Amoeba* (Rhizopoda), Figure 2-7

Amoebae are freshwater protozoans that prey on bacteria and smaller protozoans. They move by means of pseudopodia and are therefore placed in the category Rhizopoda. The organisms look like blobs of gelatin and are sometimes hard to distinguish from surrounding debris. Look for debris that appears bluish or steel gray, it's probably an *Amoeba* (actual debris is yellowish).

Although simple in form, *Amoeba* is well adapted to its environment and possesses highly structured and coordinated organelles. This relationship of form to function is readily apparent in the patterns of behavior exhibited by *Amoeba*. As you continue with the exercises keep in mind how form and function are related.

Procedure

Make a wet mount of *Amoeba* using a flat microscope slide and a coverglass. Your lab instructor will dispense the amoebae. Observe under all dry magnifications and identify as many structures as possible with the aid of Figure 2-7. If you have difficulty locating an *Amoeba*, ask your lab instructor for assistance—do not hesitate!

Estimate the size of a specimen and record the dimensions on Figure 2-7. Notice that one advancing pseudopod is dominant over the other pseudopodia. Does this pseudopod always remain dominant or does it relinquish its dominance to other pseudopodia?

Carefully observe the behavior of the cytoplasm in an advancing pseudopod. What changes do you observe in the different regions of the cytoplasm? What region of the cytoplasm undergoes the greatest rate of flow? As the pseudopodia advance, what happens at the trailing portion of the cell?

The formation of pseudopodia is possible because the microtubules and microfilaments of the cytoskeleton can assemble and reassemble rapidly in different parts of the cell. In addition, the microtubules of the cytoskeleton act as rails to transport cell organelles and materials throughout the cytoplasm.

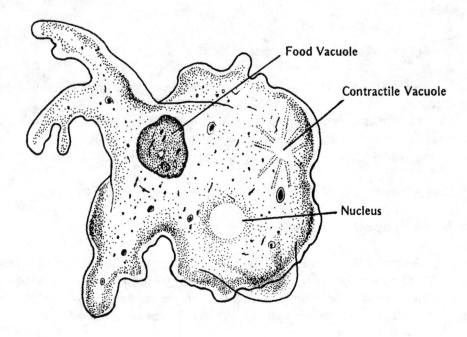

FIGURE 2-7. Amoeba

After you have identified as many structures as possible and observed the behavior of an *Amoeba*, add a small drop of methyl green to the edge of your coverglass. Gradually let the stain diffuse from the coverglass edge, or draw the stain across the slide by applying a paper towel or tissue paper to the opposite side of the coverglass. The dye will kill the *Amoeba* and stain the nucleus light green and food vacuoles blue-green.

Now prepare a depression slide of *Amoeba* that have been feeding on *Euglena*. See if you can locate an *Amoeba* that has ingested a *Euglena* or one that is capturing and engulfing its prey. This process of engulfing large objects is called **phagocytosis**. If the engulfed material is liquid or extremely small particles, the process is called **pinocytosis**.

2. *Paramecium*, Figure 2-8 and *Blepharisma* (Ciliophora), Figure 2-9

Paramecium and *Blepharisma* are freshwater protozoans placed in the phylum Ciliophora because they possess cilia. They are both predators that feed on bacteria and small protozoans. *Blepharisma* is cannibalistic and can grow to be very large. They often appear red or pink to the naked eye, and may contain many deep red food vacuoles of ingested, smaller *Blepharisma*.

Procedure

Prepare regular wet mounts of both *Paramecium* and *Blepharisma*. Observe their behavior at dry magnifications. At 40X and 100X, observe how they move through the water medium. Can they change direction? Back up? Spin? Contrast the rate of movement between *Amoeba* and *Paramecium*. Which method of locomotion do you think requires the greatest expenditure of energy per unit mass? Why? Explain.

Do the cilia appear to act independently or are they synchronized in their activity?

To slow *Paramecium* and *Blepharisma* to an observable rate at 400X, mix a small drop of methyl cellulose to a small drop of protozoan culture on a clean slide. The methyl cellulose makes the water medium more viscous. Observe as much detail as possible at 400X. Notice that the cilia beat in a synchronized pattern. Locate the oral groove of *Paramecium*. Is it lined with cilia? Why? Observe the movement of particles in the water near the oral groove. Describe this movement.

Living in freshwater presents a problem for most protozoans. The concentration of water is usually greater on the outside of their cells. Therefore, water has the tendency to move through the plasma membranes into their cytoplasm where the water concentration is less. If there were no method of voiding (pumping out) this constant inflow of water, the cell would quickly rupture. *Paramecium*, *Blepharisma*, and most other protozoans have solved this problem with **contractile vacuoles**. These vacuoles gradually fill and expel their contents through the plasma membrane into the surrounding medium. This process is not free, but costs the organism some of its chemical bond energy. These vacuoles will appear as small colorless spheres that gradually increase in size. At some stage in the filling of the vacuole you may see **radiating canals** that extend from the central portion of the vacuole. These canals bring water to the vacuole. Expulsion of the water is rapid when the vacuole is full, so you have to look constantly at a filling vacuole or you will miss it.

Make a new slide and add a small drop of methyl green to the *Paramecium*. Mix with a toothpick and add a coverslip. The dye should stain the nuclei and cause the organism's **trichocysts** to emerge from the cell surface. Trichocysts are fiber-like structures thought to have an adhesive function. They may provide *Paramecium* with a means of attachment to debris while feeding. In nonirritated cells only a few trichocysts are normally discharged, but in disturbed cells all trichocysts are discharged. Compared with cilia, how long are the trichocyst fibers?

Paramecium has two nuclei. The **micronucleus** contains the normal complement of genes and undergoes normal mitosis (division of the genetic material in the nucleus). A larger nucleus called the **macronucleus** contains multiple copies of the genes. This nucleus divides by a process other than mitosis. The primary function of the macronucleus is the regulation of cell metabolism. Both nuclei occupy a central position in the cell.

Prepare a depression slide with a small drop of *Paramecium* culture plus a small drop of red dyed yeast culture. Mix the two cultures with a toothpick, add a coverslip, and view under all magnifications. You may have to wait a few minutes, but eventually the *Paramecium* should begin to feed on (ingest) the dyed yeast cells. *Paramecium* ingest organisms by initially moving the food via the oral groove to a dead end inpocketing of the oral groove called the cytopharynx. The food item is then packaged by the cytopharynx into a vacuole and dumped into the cytoplasm of the *Paramecium*. Therefore, ingested yeast should appear as red food vacuoles. What are the length and width of your specimen in micrometers?

Length: __213.3__ µm; width: __85__ µm

(blepharisma)

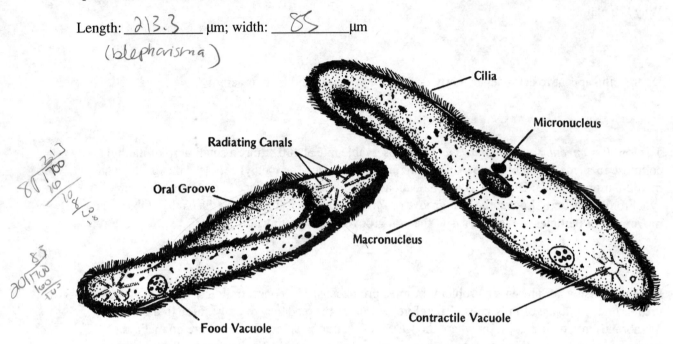

FIGURE 2-8. *Paramecium*

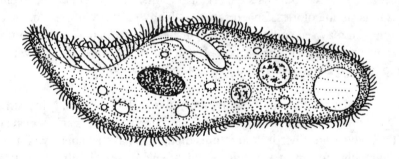

FIGURE 2-9. *Blepharisma*

III. FUNGI—Yeast (Ascomycota) and Bread Mold (Zygomycota)

This kingdom contains eukaryotic, multicellular or multinucleated organisms that obtain their energy by absorbing food from their environment (saprobes). They secrete digestive enzymes into their surrounding environment, and absorb some digested substances. Along with some bacteria, they make up the group called the decomposers. The group contains many familiar types: mushrooms, bread molds, and yeasts. Some cause diseases such as athlete's foot in humans, and Dutch elm disease in American elm. Others are widely used in our food industry. For example, yeast is widely used for baking and brewing, and many famous cheeses (Roquefort, Brie, and Camembert) derive their characteristic flavors from specific fungi that are part of the production process. A widely used antibiotic, penicillin, is produced from the fungus, *Penicillium notatum*.

Most fungi begin as spores that germinate and develop into a thread-like structure called a **hypha**. The hypha grows rapidly producing a mass of tangled hypha called a **mycelium**. Some species of hypha are **coenocytic**, having many nuclei in the same cytoplasm. Other have hypha divided by septa. The cell walls of fungi are rigid and composed of either cellulose or chitin, although the latter is the most common.

A. Yeast (*Saccharomyces cerevisiae*), Figure 2-10

The unicellular yeast *Saccharomyces cerevisiae* is the foundation of the baking and brewing industry throughout the world. In the production of wine and beer, the yeast metabolizes sugar into ethyl alcohol. In baking, a different strain of *Saccharomyces* is used to produce carbon dioxide gas that forms bubbles in the bread dough, giving the bread a light texture. Yeast can live under conditions when oxygen is either present (aerobic) or absent (anaerobic). It is classified as an Ascomycete because it produces sexual spores in saclike structures called **asci** (singular, ascus).

Procedure

Prepare a wet mount of undyed yeast. Thoroughly stir the container before dispensing a drop to your microscope slide. View under all dry magnifications. Do you see any newly forming yeast cells budding from the surface of a larger yeast cell? Prepare a new slide with red dyed yeast, and view at 400X. Can you see any additional cell structures that were not visible in the undyed culture?

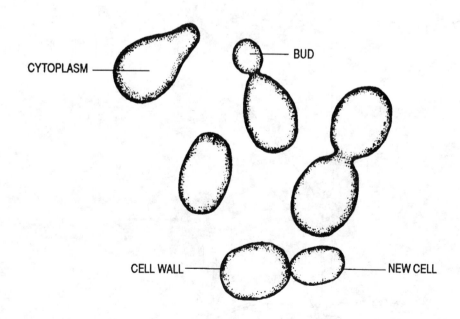

FIGURE 2-10. Yeast (*Saccharomyces cerevisiae*)

B. *Rhizopus stolonifer* (Black Bread Mold), Figure 2-11

Rhizopus looks like what we typically think a fungus should look like, that is, it consists of a mass of branching hyphae called **stolons** that grow across the surface of the bread. From these stolons grow short branching hyphae called **rhizoids** that penetrate the bread. These rhizoids are the main absorbing structures of the bread mold. When the bread mold reproduces by producing asexual spores, it produces stalk-like structures called **sporangiophores** that bear at their ends the spore producing structures called **sporangia.** The spores are dispersed to new locations by wind currents. When a spore lands upon a suitable substrate, such as bread, the spore germinates to produce new hyphae. *Rhizopus* is classified as a Zygomycete because it produces resistant structures called **zygosporangia** during sexual reproduction.

Procedure

Before you prepare your microscope slide, observe the mold under a stereomicroscope. Notice how the stolons grow across the surface of the bread, and the rhizoids grow into the bread. With tweezers, remove a small quantity of mycelium from the black bread mold culture dish and place it on a microscope slide. Add a small drop of water and cover with a coverglass. Be careful not to place too much of the mycelium on the slide. View under all dry objective magnifications. Notice that the mycelium looks white, but that mature sporangia appear black. The name black bread mold comes from the appearance of these black sporangia.

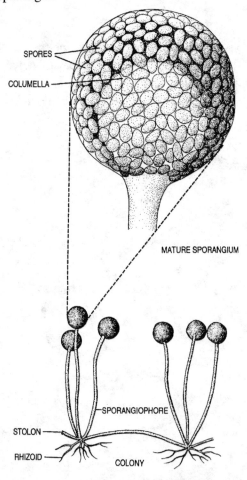

FIGURE 2-11. *Rhizopus stolonifer*

IV. ANIMALIA—Annelida and Arthropoda

Organisms within this kingdom are multicellular heterotrophs that ingest their food rather than absorb small organic molecules from their surroundings, as do the fungi. The kingdom contains the largest number of described species, approximately one million, divided among over 30 phyla. These animals live in every conceivable type of aquatic and terrestrial habitat. A single phylum, Arthropoda, contains three times the number of species than all the other phyla in the kingdom combined. A single class, Insecta, within the phylum arthropoda, contains over 700,000 species. Most people probably think of animals as frogs, fish, snakes, birds, and mammals, but less than 5% of all animals belong to this group called the vertebrates.

You will look at three microscopic animal types; a freshwater segmented worm, *Aeolosoma*, and two crustacea, *Daphnia* and *Cyclops*.

1. *Aeolosoma*—Aquatic Worm, Figure 2-12

Aeolosoma belongs to the phylum Annelida that includes the aquatic earthworms, leeches, and polychaetes. The general name for this groups is the *segmented worms*. These segmented worms are built on a tube-within-a-tube plan. A digestive tract runs the length of the body and has an anterior mouth and a posterior anus. The body wall is soft and muscular with a thin cuticle covering. The digestive tract is supported by transverse septa that mark the internal segmentation. Bristles, called setae, are arranged in bundles, two dorsolateral and two ventrolateral. The bundles of setae emerge at the septa at most segments except the first anterior segment. There are well developed closed circulatory and excretory systems.

The aquatic earthworms are common in mud and debris of stagnant pools, ponds, lakes, and streams. *Aeolosoma* feeds on debris and microorganisms swept toward the mouth by ciliary action.

Procedure

You can observe the worms in their culture dish with the stereomicroscope. Can you determine which end is anterior and which is posterior? You should be able to see the digestive tract in the transparent body of the worm. Your instructor will dispense a specimen or two of *Aeolosoma* to you in a depression slide. Cover with a coverglass and view under 40X and 100X. Can you see the chitinous setae? Are any internal organs visible?

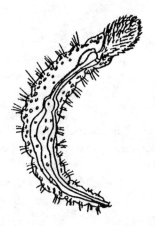

FIGURE 2-12. *Aeolosoma*

2. *Daphnia*—Water Fleas, Figure 2-13

Daphnia and the next organism you will look at, *Cyclops*, belong to the class Crustacea within the phylum Arthropoda. Most members of the Crustacea are aquatic, have two pairs of antennae, bear paired, jointed appendages at the body segments, and extract oxygen from water for respiration through gills or the body surface. There are about 30,000 described species within the Crustacea, most of which are marine.

Daphnia belongs to the order Cladocera (water fleas) within the Crustacea. They are primarily freshwater organisms, abundant everywhere except for polluted waters and fast running streams and rivers. The Cladocera are an important component of aquatic food chains, being eaten by both young and adult fish. They in turn feed upon algae, protozoans, bacteria, and organic detritus. The primary organs of locomotion are the antennae, which move *Daphnia* about in "hops."

Procedure

Daphnia run between 0.2 and 3.0 mm long, so they can be quite large. Using a stereomicroscope, look at *Daphnia* in the culture dish. You should be able to see some of the general characteristics of water fleas with the stereomicroscope. Obtain a depression slide and your instructor will dispense a few specimens to your slide. Add a drop of methyl cellulose to the depression and stir. Add a coverglass and view under 40X. The bodies of the Cladocera are not clearly segmented as in the Annelida. In most species, the body is covered by a shell or carapace that looks like the bivalve shells of clams. You should be able to see the antennae, compound eye, heart, intestine, and brood chamber.

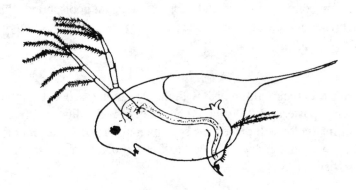

FIGURE 2-13. *Daphnia*

3. *Cyclops* —A Copepod, Figure 2-14

Cyclops belongs to the subclass Copepoda. Like the Cladocera, the Copepoda are found mostly in freshwater environments, except fast-moving waters. Many feed on freshwater algae, zooplankton, and organic detritus, while others are parasitic on fish. The free-living copepods are an important part of aquatic food chains. They form an intermediate trophic level between algae, bacteria and protozoans and larger plankton predators such as fish. *Cyclops* is an intermediate host to the fish tapeworm of humans.

Procedure

View *Cyclops* in its culture dish with the aid of the stereomicroscope. Notice that segmentation is evident along the long axis of the organism. This segmentation can be organized into three general body regions, the head, thorax, and abdomen. Some females may have bright red egg masses attached to either side of the last thoracic (genital) segment. Place one or more individuals in a depression slide and add a drop of methyl cellulose. Stir and add a coverglass. View under dry magnifications with your regular microscope. Can you see the single anterior eye spot from which cyclops derives its name? Thoracic segments 2–6 have leg appendages. Copepods are good swimmers, using both antennae and their legs for locomotion.

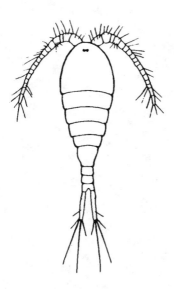

FIGURE 2-14. *Cyclops*

V. Pond Water

If time permits, you may wish to look at a sample of pond water. While looking at the sample you may see some of the same organisms you saw in the previous exercises, but you will also see some new ones. You are not responsible for identifying specific specimens nor their structures. This culture is provided so you can leisurely scan its contents, and hopefully begin to appreciate the diversity of microscopic life. So, sit back and relax as you peer into the microcosm of microscopic organisms, and enjoy the wealth of sophisticated adaptations that abound in this unfamiliar miniature world. Various guides are available to assist you in the identification of individual specimens.

Procedure

Prepare a regular or depression wet mount, whichever you prefer, and view under all dry magnifications. You may want to add a small amount of methyl cellulose to help slow the movement of the organisms in culture. You may also want to add small amounts of dye (either methyl green or methyl blue) to your slide. Feel free to experiment in preparing your slides.

Diffusion and Osmosis

Energy: chemical, mechanical, electrical, radiant, heat
(sugar) (spring, pump) (lights) (stove)
(heart)

INTRODUCTION

Anything that has mass and occupies space is called **matter**. Energy is defined indirectly in terms of the movement of matter through a distance, i.e., **work**. A change in the position or motion of matter can only come about if work is done on an object. Work is defined in terms of the force necessary to move matter across a distance, that is, Work = Force x Distance. This principle should be intuitively obvious to most of you. For example, to move a heavy object half way across a room would require a certain amount of work. You would not expect to move the same object entirely across the room by doing the same amount of work that was necessary to move the object only half way across the room. In the same vein, if you had two objects, one having twice the mass as the other, you'd need to do twice as much work on the heaviest object if you moved the objects across the same distance.

Energy associated with the movement of matter is called **kinetic** energy. Energy that exists in an inactive or stored state is called **potential** energy. Besides the two states of energy, energy may exist in many forms; heat, light, electrical, mechanical, and chemical. The first law of thermodynamics states that the total amount of energy in a system and its surroundings remains constant. That is, the amount originally present is never reduced nor increased. This implies that energy is neither created nor consumed in any chemical or physical process. Then what does occur during these processes?

When a chemical or physical process takes place, the energy of the system, and/or surroundings undergoes a transformation from one form to another (e.g., chemical energy transformed into heat energy). If all of the energy in its various forms are added after some change, the sum will equal what was originally present. No energy has been consumed. The energy has only changed its form.

To explain energy changes a few terms are necessary. The collection of matter under investigation is called the **system**, and everything outside the system is called the surroundings. Energy may pass in either direction between the system and surroundings. Thermodynamics describes the energy state of the system and surroundings before some change (**initial state**) relative to the energy state of the system and surroundings after some change (**final state**). The second law of thermodynamics relates to the tendency in the universe to maximize the degree of disorder or randomness, and minimize the **free energy** content of a system. Free energy is energy that has the capacity to do work under constant temperature and pressure. The second law tells us that chemical and physical processes never occur spontaneously in a way that decreases the randomness or disorder of the universe, i.e., the system plus surroundings. **Entropy** is the term used to describe the degree of disorder in a system and/or surroundings. *(balloon) G S mostly opposites*

To apply the preceding ideas to a familiar process, refer to Figure 3-1. Notice that the diagram on the left describes the initial condition of a container that has just had some perfume molecules placed in the left end. On the right describes the final condition of the box once the perfume molecules have dispersed equally throughout the container. What can be said about the change that has taken place using some of the terminology just introduced?

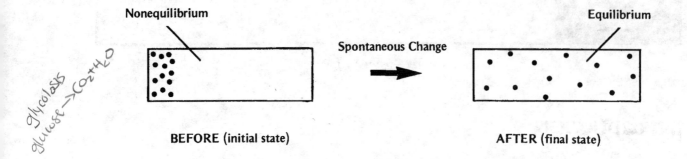

FIGURE 3-1. Diffusion of a Gas

First, because the tendency in the universe is to maximize entropy, only those movements of matter from orderly arrangements to more random arrangements can occur spontaneously. A spontaneous movement is one that will occur without an input of free energy from the outside (surroundings). Therefore, as experience has probably taught you, it is not expected once the perfume molecules have diffused uniformly throughout a volume that they would spontaneously reverse their dispersion pattern and return to a concentrated arrangement. The only way that we expect to see a random collection of matter move into a highly organized state is if work is done on the matter. In order for work to be done, free energy must be expended. Remember, free energy is that portion of the energy within a system that is available to do work. Therefore, whenever a system undergoes a spontaneous change, the amount of free energy of the system is decreased. It must decrease or no spontaneous change can occur.

By now some of you are probably wondering what all of this has to do with biology. Since the general tendency in the universe is to produce random arrangements of matter, whenever an orderly collection of matter is encountered, such as an organism, its order can be considered improbable in comparison to a more random (probable) arrangement of the same matter, Figure 3-2. How can an organism originate, maintain, and eventually perpetuate itself in such a universe? The answer is simply that organisms build and maintain their molecular configurations at the expense of the surroundings. That is, they extract free energy from the surroundings, and use it to counteract the tendency of the universe that is trying to randomize their atoms and molecules. In the process the surroundings become more disorganized. When an organism loses the capacity to extract free energy from the environment (feed) it loses the battle against increasing entropy, and its molecular structure starts to disassemble (death). The person who said "ashes to ashes" had greater insight than they probably realized.

Therefore, it's not surprising that much of the life of an organism is spent in activities that procure sources of free energy from its environment. Not all exchanges of materials between organisms and their environment involve the expenditure of free energy by the organisms. Because organisms are ordered concentrations of chemicals in an aqueous solution surrounded by a membrane that selectively allows exchange of materials, many substances may be exchanged by the spontaneous movement of matter. It is the movement of matter from more orderly arrangements (high free energy, low entropy) to more randomized arrangements (lower free energy, higher entropy) that is of principal concern in the exercise today.

The movement of particles spontaneously from a state of higher free energy to a state of lower free energy is called **diffusion**. In most biological systems this movement of particles can be interpreted as movement from a region of higher concentration to a region of lower concentration, but it more appropriately should be explained in terms of free energy or entropy. To put it another way, it is the tendency to maximize entropy and minimize free energy that is really the driving force behind the process of diffusion.

Osmosis - semi permeable membrane

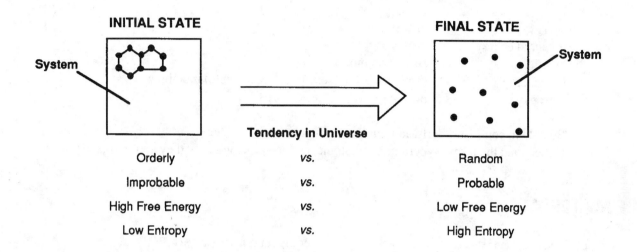

FIGURE 3-2. Arrangements of Matter

The removal or incorporation of materials by this spontaneous process is critical to the lives of most organisms. For example, the exchange of oxygen and carbon dioxide in your lungs takes place by this nonenergy requiring process. Oxygen molecules move into your blood stream because the concentration is greater in the air than in your blood stream. It then moves from your blood stream into the cells of your body where it is consumed in metabolic reactions. Because oxygen is consumed almost immediately, its concentration in the cells remains extremely low. Similarly, carbon dioxide (a metabolic waste product) accumulates in the tissues of your body, and then diffuses into your blood stream, again because of the lower concentration. Next, the carbon dioxide moves across the surface of your lungs to be released to the atmosphere. The complete exchange of these gases requires no expenditure of energy by the organism.

Sometimes the movement of substances into or out of cells by spontaneous processes are a hindrance rather than an immediate benefit. Therefore, organisms often have elaborate structures and/or processes that help eliminate excess materials that accumulate inside their cells, or that help retain substances that have the tendency to move out. For example, in a previous exercise you looked at the protozoan *Paramecium*. Because *Paramecium* is a freshwater organism, the particles in solution in the cytoplasm are at a higher concentration than the surroundings. This in turn means that the concentration of water is higher in the surrounding environment than inside the cell. Therefore, water has the tendency to constantly move through the plasma membrane and into the cytoplasm. Obviously this process cannot continue indefinitely. To prevent the cell from swelling and rupturing, the organism employs a structure called the **contractile vacuole**. This organelle acts as a pump that can evacuate the cytoplasm of excess water. To reverse the results of this spontaneous process requires the expenditure of some cellular free energy. This pumping process is not free. In an exercise that is to follow, you will investigate how freshwater algae solve the problem of the potential influx of water due to diffusion.

OBJECTIVES

1. Be able to define and apply the following terms:

energy	diffusion
potential energy	osmosis
kinetic energy	Brownian motion
free energy	semipermeable (selectively permeable) membrane
entropy	plasmolysis

2. Be able to contrast orderly and random arrangements of matter.
3. Understand the terms hypertonic, hypotonic, and isotonic and how they apply to osmosis.

MATERIALS

Living specimens:
  Closterium sp.
  Spirogyra sp.
  Paramecium sp.
carmine dye suspension
gas diffusion (demo):
  glass tube
  ring stand
  clamp (Buret)
  cotton, cork plugs
  grease pencil
  meter stick
  long test tube brush
  concentrated HCl
  concentrated NH_4OH
  microscope slides and coverslips

compound microscopes
distilled water
liquid diffusion (demo):
  $KMnO_4$ crystals
  three–100 ml graduated cylinders
Osmosis:
  buret clamps
  ring stand
  large thistle tubes
  400 ml beakers
  animal membrane and rubber band
  30% sugar solution (dyed green)
  lens paper
  lens cleaning solvent
  toothpicks
  NaCl solutions

I. DIFFUSION

A. Brownian Motion—The Random Movement of Microscopic Particles Suspended in a Gas or Liquid

1. Brownian Motion in *Closterium* Vacuoles (Student Preparation)

Obtain a small drop of *Closterium* (a unicellular green alga) from the culture dish and prepare a wet mount. A vacuole containing calcium sulfate crystals (gypsum) can be found at each end of this crescent shaped alga.

Notice that these crystals are in constant, erratic motion. Why?

 HINT: These crystals are suspended in a fluid environment that consists of much smaller particles. Remember that all atoms, molecules, and ions in any system are in a constant state of random motion.

2. Brownian Motion in a Carmine Dye Suspension (Student Preparation)

Prepare a wet mount with carmine dye suspension. Shake the dropper bottle before placing a drop of dye on your microscope slide. Add a coverglass and view under 400X. Remember to first find the particles at 40X than 100X, and finally at 400X. Look carefully because the particles in this dye suspension are much smaller than the large crystals found in *Closterium*. Do not touch your scope excessively or bump the table while you are looking for Brownian Motion. If you have difficulty seeing the movement of the dye particles adjust the iris diaphragm and reduce the light intensity. In the space below, diagrammatically explain how Brownian Motion occurs.

B. Diffusion of a Gas through a Gas (Instructor and Student Demonstration)

$$\frac{r_A}{r_B} = \frac{\sqrt{\text{molecular weight}_B}}{\sqrt{m_A}}$$

Your instructor and one student will conduct the experiment. Two cotton plugs will be inserted simultaneously in opposite ends of a horizontal glass tube. One plug will be saturated with concentrated hydrochloric acid and the other with ammonium hydroxide. Put the same number of drops of liquid on both plugs. The diffusing molecules of principal concern are hydrogen chloride (HCl) and ammonia (NH_3). When HCl and NH_3 molecules react they form a white solid, ammonium chloride (NH_4Cl). Note: The molecular weights of NH_3 and HCl are 17 and 36.5, respectively.

With a grease pencil, mark the spot where the white precipitate initially appears. Measure the distance that each type of molecule diffused. Do the distances suggest anything about the relative rates at which the gas molecules diffused? Explain.

List some factors that could influence the rate at which these molecules diffuse.

How could you determine whether these factors influence the rate of diffusing gases?

> ☞ **HINT**: Is there any way you could adapt this experiment to test any of these potential variables? Explain.

C. Diffusion of a Solid through a Liquid (on Demonstration)

Crystals of potassium permanganate ($KMnO_4$) were placed on the bottom of a series of graduated cylinders filled with water. The date and time that each cylinder was started is recorded at the base of the individual cylinders. Notice that the rate of diffusion in this experiment is much slower than the diffusion rate of the preceding experiment. Why?

> **CAUTION**: Do not disturb the graduated cylinders, as the slightest disruption will upset the random diffusion of particles.

II. OSMOSIS—A SPECIAL TYPE OF DIFFUSION

Osmosis is the diffusion of a solvent, usually water, through a **semipermeable** (selectively permeable) **membrane** from a region of greater solvent concentration to one of lesser concentration. More appropriately, the net movement of water is from a state of greater free energy and lower entropy to a state of lower free energy and higher entropy. The driving force behind the process of osmosis is the same as diffusion, that is, the tendency of a system to spontaneously reduce its free energy content and maximize its disorder (entropy).

Consider for a moment a container (system) filled with distilled water (Figure 1-3a). the container is divided in half by a semipermeable membrane that only allows water molecules to freely pass through it; all other molecules are repelled by the membrane. In the situation just outlined there is no **net** movement of water to either side of the container, although water molecules can still be exchanged from one side to the other at the same rate. For example, for every ten molecules that move from the left side to the right, there must be ten molecules that move in the reverse direction. Notice that the system is at a uniform concentration throughout, and therefore cannot undergo a spontaneous change. This is analogous to having a container at uniform temperature. Once the container is at the same temperature as the surroundings, it cannot spontaneously get hotter or colder.

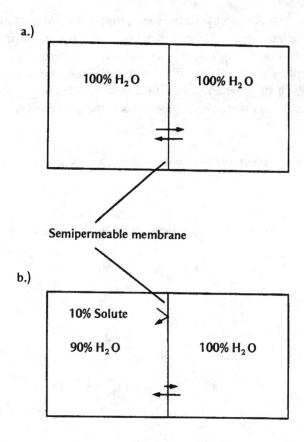

FIGURE 1-3. Osmosis

Next consider the same container of water, but on the left side 10% of the volume of water is replaced by particles dissolved in solution (Figure 1-3b). Can there be a spontaneous change in the system just described? Let us first consider the particles in solution. It should be evident that the particles would diffuse to the right side of the container if they could, but they are restricted from doing so because of the membrane. Therefore, the particles remain localized on the left side of the container, and no gross nor net movement of the particles through the membrane occurs. Is the same true of the water molecules? To understand how water will respond in this system it will be helpful if you can ignore the kinds and amounts of particles in solution, and restrict your attention to the relative concentrations of water throughout the system. For example, in the situation outlined above the left side of the container has 90% water while the right side has 100%. The only spontaneous change that can occur is the movement of water from the area of greater concentration to lesser concentration. Therefore, the rate of movement of water is greatest from the right to the left side.

A. An Osmotic Apparatus (on Demonstration), Figure 3-4

Solutions may be classified as **hypertonic, hypotonic**, or **isotonic** depending on the number of solute particles in solution relative to the number of solute particles in another solution. A solution that has more dissolved solute particles per unit volume, relative to another solution, is hypertonic. A solution with fewer dissolved particles per unit volume is hypotonic, and a solution with the same number of dissolved particles per unit volume is isotonic. You can use these descriptions to predict the relative concentrations of water in two solutions, and if the solutions are separated by a semipermeable membrane you can predict the direction of any net movement of water.

Prior to lab, two thistle tubes were set up in your lab room, Figure 3-4. Animal membranes were attached to the wide end of each thistle tube. These membranes act as semipermeable membranes, preventing substances from passing through them, such as sucrose (table sugar), but allowing water to freely flow through them. Thistle tube no. 1 was filled with a 30% sucrose solution, and then placed in a beaker of distilled water. Thistle tube no. 2 was filled with distilled water, and then placed in a beaker containing a 30% sucrose solution. The 30% sucrose solution is dyed with green food coloring.

Is thistle tube no. 1 hypertonic, hypotonic, or isotonic to the beaker solution? Explain.

Is thistle tube no. 2 hypertonic, hypotonic, or isotonic to the beaker solution? Explain.

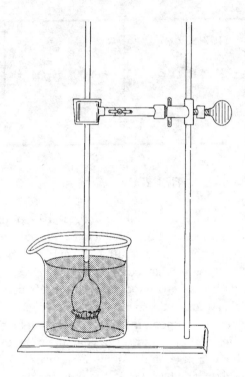

FIGURE 3-4. Thistle Tube Apparatus

B. Osmosis in *Paramecium*—Measuring the Contractile Vacuole Expulsion Rate

Because all freshwater protozoans have a higher solute concentration within their plasma membranes than their surrounding environment, they have a constant influx of water into their cells. Therefore, they must continually pump excess water back into the surrounding or their cells will swell and rupture. *Paramecium* and other freshwater protozoans solve this problem using a pumping structure called the contractile vacuole to rid their cells of excess water. You might not think that the rate of water influx into the cells of *Paramecium* and other freshwater protozoans is very fast. I hope that after you complete the following exercise, you will appreciate just how much water they must expel compared to their size.

In the following experiment, you will determine the rate of water evacuation from the contractile vacuoles of *Paramecium* by measuring the pulsation rate (pulsations/min.) of the vacuoles.

Procedure

1. Add a small drop of methyl cellulose to a clean slide. Then add a small drop of *Paramecium* from the regular culture dish and mix with a toothpick. The methyl cellulose does not alter the osmolarity of the solution, but will slow the movement of the *Paramecium* so you can make your observations. Add a coverglass and locate an individual *Paramecium* that has at least one of its two contractile vacuoles visible. The anterior vacuole is probably the best to use because it is the least obscured. Probably the best magnification for making your counts is 100X. Make three separate observations for two minute each, counting the number of times the vacuole empties (pulses). Compute an average for the three readings and record this average in Table 3-1.

Table 3-1. Contractile Vacuole Evacuation Rates				
Environmental Medium	**Evacuation Rate (pulsations/min.)**	**Vac. Vol. (μm^3)**	**Vac. Output/min. (μm^3/min.)**	**Total Output (μm^3/min.)**
regular culture	3	4188	12,504	25128
1.0% sorbitol				

To compute the total volume of output per minute of a contractile vacuole, you would need to compute the approximate volume of the vacuole. Assuming a vacuole is roughly spherical and has a radius of approximately 10 µm, you would use the following formula for the volume of a sphere:

$$\frac{Pulse}{min} \times \frac{vol}{pulse} = \frac{vol\ flow}{min\ vac}$$

$$Volume = \frac{4\pi r^3}{3} \qquad \text{where } \pi = 3.1416$$

To compute the total output of the vacuole per minute, you would simply multiply the volume of the vacuole times the number of pulses per minute. Calculate the total output per minute for the entire organism, remember that *Paramecium* has two contractile vacuoles, one anterior and the other posterior, enter your results in Table 3-1. What calculations would you have to make to determine what percentage of the volume of a *Paramecium* must be evacuated every minute? Hint: The formula for calculating the volume of a circular, right angle cylinder, is: $\pi r^2 h$.

2. If time permits, repeat the procedure above, but use *Paramecium* from a 1.0% sorbitol solution. The plasma membrane of *Paramecium* is not permeable to this sugar alcohol. Locate a specimen and make three separate observations for 2 minute each. Compute an average for the three readings and record this average in Table 3-1.

Is the observed rate of evacuation higher or lower than the rate for *Paramecium* in freshwater? Is this what you expected? If not, how do you account for the observed rate?

> ☞ **HINT**: What is the diameter of each vacuole in the sorbitol solution?

III. PLASMOLYSIS (OSMOSIS IN ALGAL CELLS)

In the preceding experiment you investigated the osmotic process using a freshwater protozoan that had a mechanism to rid itself of excess water, the contractile vacuole. What happens in algal cells that live in freshwater environments? Their cytoplasm contains more solute particles than the surrounding freshwater, and their plasma membrane is permeable to water. They also lack contractile vacuoles. Why don't their cells swell and rupture?

Algal cells solve this apparent paradox by using a passive structure to retard the influx of water, the **cell wall**. The cell wall is pervious to water and most dissolved substances, but it is somewhat rigid and resists the outward swelling that would be due to an influx of water. As the cytoplasm and vacuoles of a cell gain water, the cell swells outward, pressing the cytoplasm against the cell wall causing what is known as osmotic pressure. At a certain point, the pressure that is being exerted against the cell wall is equal to the pressure of the water that is trying to flow into the cell due to osmosis. At this point, there is no net movement of water into the cell.

What happens if an algal cell is placed in a hypertonic solution? Because the algal cell contains more water in its cytoplasm than the surrounding environment, the algal cell will lose water. Osmotic pressure is reduced, and the cytoplasm and the plasma membrane collapse and shrink. An algal cell in this state is said to be **plasmolyzed**. If the duration and severity of plasmolysis has not been too extreme, the plasmolyzed condition of a cell is usually partially reversible.

Procedure

Prepare a wet mount of *Spirogyra* and identify the cell structures labeled in Figure 3-5. Without removing your slide from the microscope, replace the water that contains the *Spirogyra* with a 10% NaCl (salt) solution. Do this by adding the salt solution to one edge of the coverglass (using a micropipet) while simultaneously touching the opposite edge with the edge of a dry paper towel. This should remove the original water and draw the salt solution under the coverglass. What happened to the protoplasm of the cells?

Notice that the cell walls retain their original form (morphology). Is the salt solution hypertonic, hypotonic, or isotonic to the algal cells?

Remove the salt solution from your slide and replace it with distilled water. Is plasmolysis at least partially reversible? Are contractile vacuoles present in Spirogyra?
Why or why not?

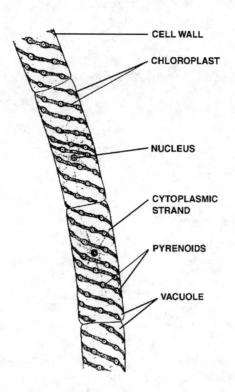

FIGURE 3-5. *Spirogyra*

4 Enzymes

INTRODUCTION

Enzymes are organic catalysts. That is, they are compounds of carbon that are responsible for accelerating the thousands of chemical reactions that constitute metabolism, and are the largest and most specialized group of proteins. Without enzymes, chemical reactions would occur at slow or imperceptible rates. If this were the case, not enough energy would be available to sustain the structures and functions of organisms.

Some enzyme molecules can catalyze reactions at phenomenal rates, while others work much slower. For example, the following reaction is catalyzed by carbonic anhydrase:

$$CO_2 + H_2O \longrightarrow H_2CO_3 \text{ (carbonic acid)}$$

This reaction allows for the rapid transfer of CO_2 from tissues to the blood. The CO_2 is then liberated to the atmosphere via the alveolar sacs of the lungs. This enzyme is one of the fastest known; a single enzyme molecule can catalyze the reaction 600,000 times in one second. Because the enzyme catalyzes a reaction that is crucial to a basic life process (respiration), it's not surprising that there has been strong selection over millions of years to increase the enzyme's rate. In fact, carbonic anhydrase displays kinetic perfection, that is, the enzyme is not limited by any physical, internal constraint in further increases in rate. The reaction rate is ultimately limited by the rate at which the enzyme and substrate encounter each other during normal diffusion.

The primitive form of the enzyme carbonic anhydrase most likely did not have as great a catalyzing ability as the contemporary enzyme. How can the catalytic ability of an enzyme be modified over time?

Because the functional shape of an enzyme molecule is primarily determined by the genetically specified amino acid sequences, enzymes are prime candidates for the action of natural selection. When a mutation occurs that changes the amino acid composition of a protein and alters the catalytic ability of the enzyme, the organism bearing the mutation may experience a reproductive advantage over other individuals of the same species. Through reproductive success, the mutation and altered gene product would gradually spread through the population, becoming common rather than rare. Conversely, old versions of the enzyme molecule would become increasingly scarce because of reduced reproductive success. Therefore, the primitive form of carbonic anhydrase would have gradually improved in efficiency as mutations occurred in the genetic material encoding the instructions necessary to build the molecule. Through the process of natural selection (differential reproductive success), enzyme structure and function can be modified over time.

The specific three-dimensional configuration of an enzyme molecule is essential for normal enzyme activity. Chemical and physical factors, such as temperature and pH, can affect the conformation of enzyme molecules, and therefore their activity.

How Do Enzymes Participate in Chemical Reactions?

The **collision theory** is fundamental to interpreting the activity of chemicals during reactions. The theory states that all atoms, molecules, and ions are in a constant state of motion, and that for two particles to interact chemically, that is, to exchange or rearrange electrons, they must first come into contact. A collision between two atoms can only occur if sufficient energy is present to overcome the mutual repulsion of the respective electron clouds of negativity that surround each atom.

Not all collisions result in successful chemical reactions. A chemical reaction, such as A + B → C + D occurs because at any given moment a certain number of A and B particles possess the **minimum kinetic energy** necessary to exchange or rearrange electrons (make or break chemical bonds) if they collide. Each type of chemical reaction has a specific minimum kinetic energy for reaction. The **rate** of a chemical reaction is therefore dependent upon the number of reactant molecules in a system that have sufficient minimum kinetic energy to react.

Not all particles in a system possess identical kinetic energies. Rather, it is necessary to think of the particles as existing as a distribution of different kinetic energies. Because the kinetic energy of a particle is proportional to the square of its velocity, we can think of the distribution as one of **velocity** rather than one of kinetic energy. If you consult Figure 4-1, you will see that the particles in a reaction are spread over a range of velocities. Notice that there are fewer particles with low and high velocities than there are particles with velocities somewhere between these extremes. The point where the vertical line drawn through the distribution of particles intersects the x-axis is called the **mean** or **average velocity**. Another name for average velocity is **temperature**. Also notice the point on the x-axis labeled **K.E.***. This is the minimum kinetic energy necessary for a reaction to occur between the reactant molecules in this example. All the molecules to the right of the K.E.*, those in the shaded area, have sufficient velocities to react.

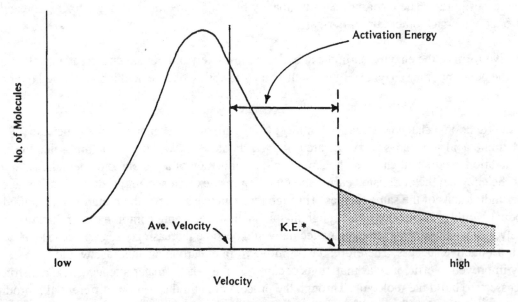

FIGURE 4-1. Distribution of Kinetic Energies and Activation Energy

The next major point to be illustrated, by Figure 4-1, is that of **activation energy**. Activation energy is the amount of kinetic energy or velocity necessary to increase particles from the mean velocity to the minimum kinetic energy for reaction. The amount of activation energy necessary for reaction can greatly influence a reaction's rate.

It follows from the discussion so far, one way to increase the rate of a reaction is to increase the number of particles that have the minimum kinetic energy for reaction. One way to do this would be to **heat** the system so a larger proportion of the molecules have the minimum kinetic energy for reaction, i.e., a large proportion occur in the shaded area of the diagram, Figure 4-2. Effectively, you would be promoting molecules up to the minimum kinetic energy for reaction.

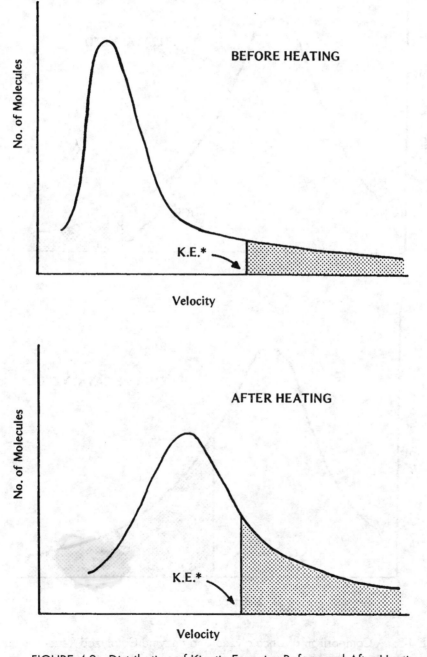

FIGURE 4-2. Distribution of Kinetic Energies Before and After Heating

Exercise 4 — Enzymes   49

Because of its deleterious effects, heating biological systems has limited value for increasing the rate of chemical reactions. Rather, another way to increase the rate would be to somehow **lower** the minimum kinetic energy for reaction, and therefore decrease the amount of activation energy necessary for reaction, Figure 4-3. This would have the effect of including more molecules in the distribution with sufficient minimum kinetic energy for reaction. This is exactly what enzymes do to accelerate reactions. By bonding to substrate molecules, they produce a new minimum kinetic energy for reaction that is lower than the uncatalyzed reaction.

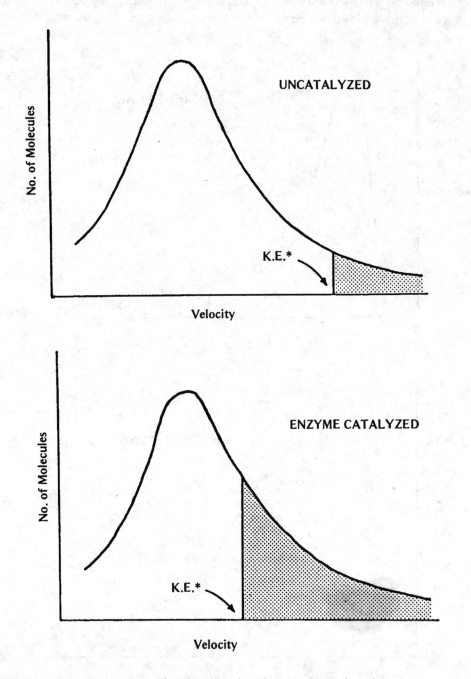

FIGURE 4-3. A Comparison of Uncatalyzed and Enzyme Catalyzed Reactions and the Minimum Kinetic Energy for Reaction

Most chemical reactions are **reversible**, $A + B \rightleftharpoons C + D$. Can the collision theory help to explain this observation? Imagine within a system there is a large concentration of the reactants A and B, and a small concentration of the products C and D. Initially, the number of collisions between A and B is very high, but as the reactants are converted to the products, C and D, the probability of successful collisions tapers off. As the reactant concentration is decreasing, the concentration of products is increasing. The products, like the reactants, are in a constant state of motion and are colliding randomly. At any one instant, a certain fraction of the product molecules may possess adequate minimum kinetic energy to form reactant molecules. When the rate of conversion between reactants and products is the same, the system is in **chemical equilibrium**. Note that chemical equilibrium is a statement about **equal rates** and not about equal concentrations. At chemical equilibrium, the concentration of the reactants and products may be greatly different if there is a large difference between the free energy of the reactants and products.

In a reaction where the products are at a much lower free energy than the reactants, a large energy barrier must be overcome to drive the reverse uphill reaction. Therefore, in this type of reaction, the concentration of products must increase to a point where a portion of the product molecules possess sufficient activation energy to drive the reverse reaction. The larger the difference in free energy between the reactants and products, the larger the energy barrier that must be overcome, and therefore, the larger the difference will be in the concentration of the reactants and products. Figure 4-4 is an energy diagram for catalyzed and uncatalyzed reactions. Notice that the amount of free energy necessary for the forward uncatalyzed reaction (A) is greater than that necessary for the forward catalyzed reaction (B). The amount of free energy necessary for a reverse reaction (C) is greater than either A or B.

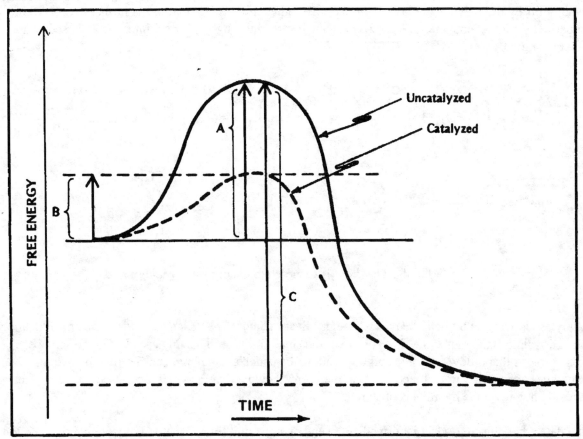

FIGURE 4-4. Energy Diagram of Catalyzed and Uncatalyzed Reactions

Exercise 4 — Enzymes

Another factor that influences the probability of collisions between particles within a system is **concentration**. As the number of particles in a system increases, the more collisions per unit time that have the minimum kinetic energy for reaction. To further increase the probability of successful collisions, enzymes provide a means by which the reactive sites of reactant molecules are oriented. If two particles collide with their reactive sites in improper alignment, the reaction may not occur. **Orientation** of reactive sites is therefore extremely important when the reacting molecules are large.

During an enzyme-catalyzed reaction, the enzyme molecule combines with the substrate to form what is called an **enzyme substrate complex (ES)**. Therefore, a single enzyme molecule can catalyze only one reaction at a time, and it can do it only so fast before it can participate in another identical catalyzed reaction. In other words, given that an excess of reactant molecules is present, an enzyme has a **maximum rate** at which it can catalyze a reaction at any given temperature, pressure, and pH.

In today's exercise you will investigate some properties of enzymes by looking at the different changes in rate during an enzyme-catalyzed reaction. You will also examine how temperature, pH, concentration of enzyme, and concentration of reactants influences the rate of a reaction. The rate of a reaction can be measured by the amount of product formed per unit time or the amount of reactants converted per unit time. Today, you will measure the amount of **product** formed per unit time.

The enzyme reaction that you will investigate is the hydrolysis of starch (amylose) to the products glucose and maltose, Figure 4-5, by the enzyme amylase. Starch is a polymer (repeating structural unit) of glucose, and hydrolysis is the breaking apart of organic compounds by the addition of water. In this hydrolysis reaction, starch is cleaved into glucose (a monosaccharide) and maltose (a disaccharide) units by the enzyme amylase. In humans and other animals, the maltose is further hydrolyzed to glucose by another enzyme. The glucose is then absorbed through the small intestine and used as an energy source.

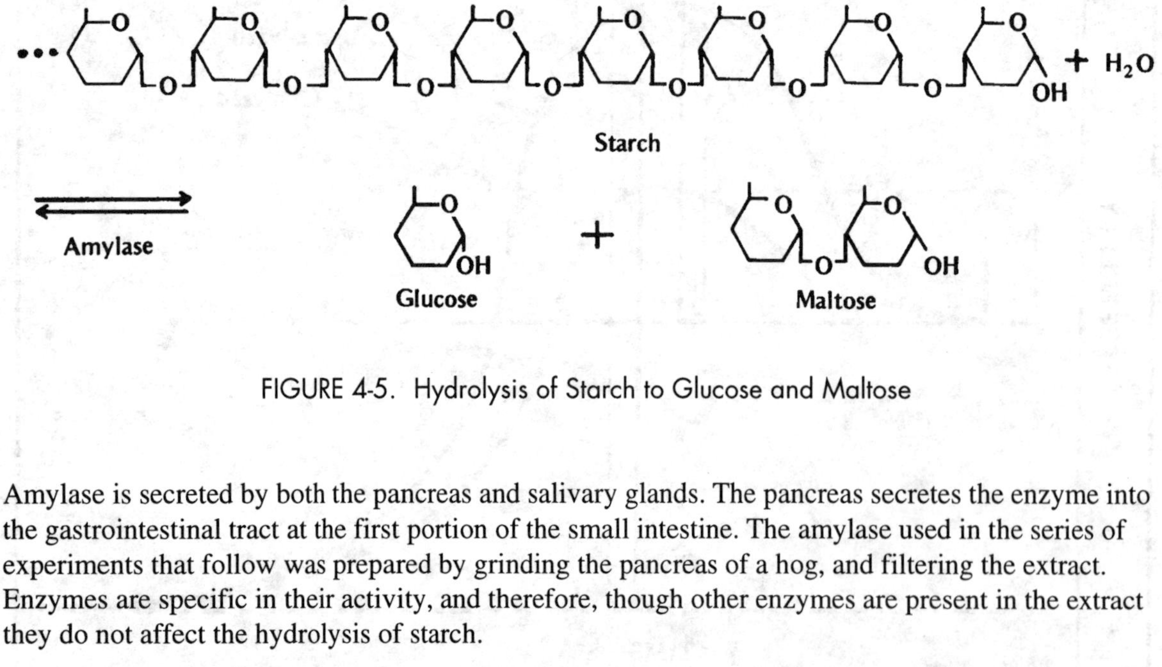

FIGURE 4-5. Hydrolysis of Starch to Glucose and Maltose

Amylase is secreted by both the pancreas and salivary glands. The pancreas secretes the enzyme into the gastrointestinal tract at the first portion of the small intestine. The amylase used in the series of experiments that follow was prepared by grinding the pancreas of a hog, and filtering the extract. Enzymes are specific in their activity, and therefore, though other enzymes are present in the extract they do not affect the hydrolysis of starch.

To detect how much product is formed at the end of some unit time, a method of stopping the reaction is necessary. In addition, a method for measuring the amount of glucose and maltose in solution is essential. A color reagent, 3,5-dinitro-2-hydroxy-benzoic acid, serves the dual function of stopping the reaction and forming a colored compound with glucose and maltose. When glucose and maltose are in an aqueous solution, they are essentially colorless, and when the color reagent is in solution alone it is yellow. After a color reagent molecule has attached to either glucose or maltose, the newly formed molecule appears amber in solution, Figure 4-6. Note: A glucose molecule, plus color reagent molecule, and a maltose molecule, plus color reagent molecule, appear the same color (amber).

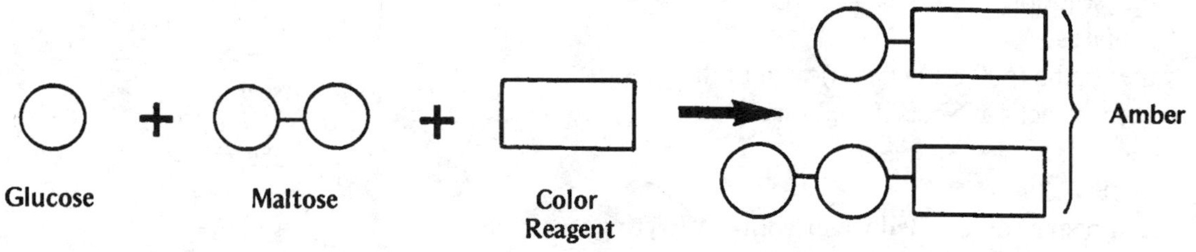

FIGURE 4-6. Abstract Representation of the Color Reagent Reaction with Glucose and Maltose

A **spectrophotometer** can then be used to measure the amount of light absorbed by the colored compound. For information about the theory and operation of the spectrophotometer, consult Appendix II. Therefore, when a solution of colored product is placed in a spectrophotometer, the machine produces an absorbance value that is proportional to the concentration of the colored compound. In turn, the concentration of colored product is proportional to the concentration of glucose and maltose in solution. Thus we have a simple technique for detecting the amount of glucose and maltose produced by the enzyme hydrolysis of starch.

OBJECTIVES

1. Define:

    activation energy
    concentration
    chemical equilibrium
    enzyme
    standard curve
    substrate
    control
    enzyme-substrate complex (ES)

2. Be able to explain how temperature, pH, enzyme concentration and substrate concentration affect the rate of chemical reactions.
3. List three factors that determine, in part, whether a collision between two particles will lead to a successful chemical reaction.
4. Explain diagrammatically how starch is hydrolyzed by amylase to glucose and maltose.
5. Be able to answer questions based on graphs produced from enzyme-catalyzed actions.
6. Be able to explain why it is necessary to produce a standard curve.
7. Be able to explain why it was necessary to do an enzyme rate experiment before measuring the effect of enzyme concentration on the reaction rate.

MATERIALS

    enzyme solution*
    color reagent solution with 1.0 ml syringe and canula*
    starched buffer solutions: pH 3,5,7,9, and 11*
    phosphate buffer, pH 7*
    glucose/maltose standard*
    NaCl solution*
    hot plates
    water baths (600 ml beakers) with boiling beads
    test tube racks at Spec 20 stations
    Spec 20's
    cuvettes
    small beaker of distilled water with 5 ml syringe
    grease pencil
    beakers (approx. 250 ml for temperature experiment)
    ice
    dropper bottles with Pasteur pipettes (for solutions listed above)
    alcohol thermometers

*Refer to the accompanying preparation manual for directions in preparing these solutions.

I. DETERMINING A STANDARD CURVE

Absorbance values alone tell you nothing about how much of a substance is in a solution, i.e., concentration. Therefore, it is necessary to generate a set of absorbance values that correspond to a series of known concentrations. By plotting these absorbance values against known concentrations you can produce a reference line called a "standard curve."

Once you have produced a standard curve, you can use it to determine which absorbance values correspond to which concentrations of product. Two important conditions must be met before you can use the standard curve in other experiments. One condition is that the **products** of the enzyme-catalyzed reaction you are investigating must be the same as those used to produce the curve. The other condition is that the **volume** of the unknown product solution must be the same as the volumes of the solutions used in producing the curve. Why are equal volumes important? Remember that concentration is an expression of the number of particles per unit volume. Therefore, when you compare concentrations, you must keep the volume constant. In other words, the number of particles, not the volume, should be the variable.

Having only one variable, with all other factors held constant, is extremely important during any experiment. Throughout each experiment remember to apply this **one variable** approach. In the procedure that follows, you will use known concentrations of a solution containing one half glucose and one half maltose to produce a standard curve. The color reagent reacts in the same manner with both glucose and maltose.

Procedure

1. Label 8 test tubes B, 1, 2, 3, 4, 5, 6 and S, respectively.
2. Prepare the tubes as follows:

Tube	Drops* of 50:50 Glucose/Maltose Solution	Drops* of Phosphate Buffer, pH 7	Drops* of 1% Starch, pH 7
B	0	14	0
1	2	12	0
2	4	10	0
3	6	8	0
4	8	6	0
5	10	4	0
6	12	2	0
S	0	0	14

*Although a drop is a relative unit of measure, it is a convenient and a reproducible volume for many experiments. Use one pipette for each reagent; do not mix pipettes. The term drop refers to a free-falling drop. Do not touch the drop to the wall of a container before it is fully formed. Hold the pipette at the same angle for each drop.

3. With a syringe, add 0.5 milliliter (ml) of color reagent to each tube and mix thoroughly by swirling, shaking or "tweaking" the tube.

 WARNING: The color reagent is caustic because it contains sodium hydroxide (lye). Wipe up any spills or splashes. If you spill any on your skin, wash immediately.

4. Heat the tubes for five minutes in a boiling water bath. The reaction between glucose/maltose solution and the color reagent is accelerated by this heating.
5. Remove the tubes from the water bath and add exactly 2.5 ml of distilled water to each tube.
6. Mix the contents of each tube thoroughly (until the color is uniform throughout the tube) and wipe off the outside of the tube.
7. Adjust the spectrophotometer to a wavelength of 540 nm, and set the meter to read zero percent transmittance (left, front knob).
8. Insert sample tube B (the blank) and close the lid of the sample holder. This sample will serve as a reference with a designated absorbance of zero. Therefore, adjust the Spec 20 to register an absorbance of zero, or 100% transmittance (right, front knob).
9. Remove the blank and do not change any instrument settings. Insert each sample tube in turn and record the absorbance values obtained in Table 4-1.
10. Plot the absorbance values as a function of the amount of glucose and maltose present in tubes 1-6 (Graph 1).

TABLE 4-1

Tube	µmoles per Tube	A₅₄₀
1	0.26	.14
2	0.52	.39
3	0.78	.37
4	1.04	.55
5	1.30	.72
6	1.56	.66
S	0	.035

Questions

1. How does applying heat to the maltose and color reagent solution speed up the reaction?

   Catalyst

2. Speculate about why you use a wavelength of 540 nm rather than another wavelength?

3. Is the standard curve a straight line? Why or why not?

   straight

4. Why did you add various amounts of phosphate buffer to the test tubes?

5. What was the purpose of tube B?

II. DETERMINING THE MAXIMUM RATE OF AN ENZYMATIC REACTION OR HOW FAST CAN AN ENZYME WORK?

To measure the maximum rate of an enzymatic reaction, you will create a situation where, initially, the number of substrate molecules is much larger than the number of enzyme molecules. Why is this important?

The rate of starch hydrolysis will be determined by setting up a series of mixtures of enzyme and starch and allowing the reaction in each mixture to proceed at room temperature for a specified time. The rate of the reaction is the amount of product (maltose and glucose) formed per unit time. You will analyze, for the amount of maltose and glucose, by the same chemical procedure used in Experiment I.

Procedure

1. Label 6 tubes B, 2, 4, 6, 8 and 10, respectively, to represent a blank (B) plus action times of 2, 4, 6, 8, and 10 minutes.
2. Add 10 drops of 1% starch (pH 7) to each tube.
3. Prepare the blank (B) by first adding 0.5 ml of color reagent and then adding 2 drops of the enzyme solution. Since the color reagent destroys amylase activity, the blank indicates the amount of product present at time zero.
4. Label a seventh tube 10' and place 10 drops of 1% starch (pH 7) in it. This tube will be used to measure the amount of starch hydrolyzed during 10 minutes, if any, without amylase.

> **WARNING:** DO NOT ADD ENZYME TO THIS TUBE!

5. Initiate the amylase reactions by adding 2 drops of enzyme solution to the appropriate tube at two minute intervals (starting with tube 10). After adding the enzyme, mix the contents of each tube thoroughly and allow the reaction to proceed at room temperature.
6. Ten minutes after initiating the first reaction, add 0.5 ml of color reagent to each tube, including the 10' tube. This stops the enzymatic activity. Then add 2 drops of enzyme to the 10' tube.
7. Mix the contents of each tube. Heat all tubes in a boiling water bath for five minutes and then remove them. Add 2.5 ml of distilled water to each tube. Mix thoroughly until the color in each tube is uniform.
8. Adjust the absorbance of the blank to zero and record the absorbance values of each sample in Table 4-2.
9. Use the standard curve from Experiment I to convert the absorbance values to the amount of product formed in each sample. Calculate the maximum rate of amylase-catalyzed reaction in terms of product produced per minute. Maximum rate = _____ μmoles/min.
10. Plot the μmoles of product formed as a function of the reaction time on Graph 2.

TABLE 4-2

Tube	Reaction Time (min)	A_{540}	μmoles Maltose + Glucose
0 = B	0	0	.26
2	2	.26	.52
4	4	.7	.78
6	6	1	1.04
8	8	1.1	1.30
10	10	1.3	1.56
10'	10	6	0

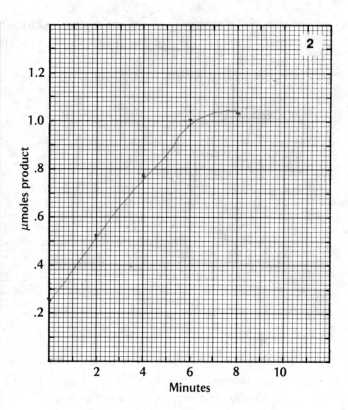

Questions

1. Why was it not necessary to add varying amounts of buffer solution to the sample tubes?

   To maintain the ph

2. If you obtained a straight line for the entire 10 minutes of reaction time, would you expect this linear relationship to continue indefinitely? How would you explain an eventual decrease in the rate of reaction? No, the reactants would be used up

3. Was there any hydrolysis of starch from the uncatalyzed reaction in the 10' tube? If not, why was it necessary to prepare this reaction tube?

   No

4. What factors were held constant throughout the experiment?

III. THE EFFECT OF SUBSTRATE CONCENTRATION ON THE REACTION RATE

When the substrate, not the efficiency of the enzyme, is the limiting factor in a reaction, the rate is proportional to the concentration of substrate. In other words, when the substrate concentration is low, the enzyme never reaches its maximum rate of conversion. Why?

> **HINT**: Probability of successful collisions.

When the substrate concentration is limiting, only then can it influence the rate of reaction. If the concentration of substrate increases, to a point where the enzyme molecules are working as fast as they can, further increases in the concentration of substrate will not affect the rate of reaction.

Examine this relationship between the reaction rate and the initial starch concentration when the amount of amylase is constant. You will calculate the rate of each reaction by determining how much product was formed during 10 minutes of reaction.

Procedure

1. Mark a set of tubes B, 1, 2, 3, 4, 5 and 6. Prepare as follows:

Tube	Drops of 1% Starch, pH 7	Drops of Phosphate Buffer, pH 7
B	10	0
1	1	9
2	2	8
3	3	7
4	5	5
5	8	2
6	10	0

2. Prepare the blank (tube B) by first adding 0.5 ml of color reagent and then adding two drops of enzyme. Mix well.
3. Initiate the reactions in all the other tubes at the same time by adding two drops of enzyme to each. Mix the contents of each tube thoroughly. Allow the reaction to proceed for ten minutes at room temperature.
   NOTE: The total volume of the 1% starch solution in each reaction mixture is 12 drops (2 drops enzyme + x drops 1% starch + x drops phosphate buffer). Therefore, the starting starch concentrations in the tubes are as follows:

Tube	Starting Starch Concentration
1	1/12 of 1% = 0.08%
2	2/12 of 1% = 0.16%
3	3/12 of 1% = 0.25%
4	5/12 of 1% = 0.41%
5	8/12 of 1% = 0.67%
6	10/12 of 1% = 0.83%

4. Stop the reaction after ten minutes by adding 0.5 ml of color reagent to each tube except the blank. Mix. Heat all tubes in a boiling water bath for five minutes. Remove.
5. Add 2.5 ml of water to each sample. Mix until the color is uniform.
6. Insert the blank and adjust the spectrophotometer to register an absorbance of zero. Measure the absorbance of each sample and record your measurements in Table 4-3.
7. In Graph 3, plot the absorbance of each sample as a function of the initial starch concentration.

Questions

1. Do all the values, plotted on Graph 3, express the maximum rate at which amylase can convert starch? Why?

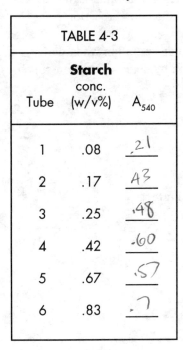

TABLE 4-3

Tube	Starch conc. (w/v%)	A_{540}
1	.08	.21
2	.17	.43
3	.25	.48
4	.42	.60
5	.67	.57
6	.83	.7

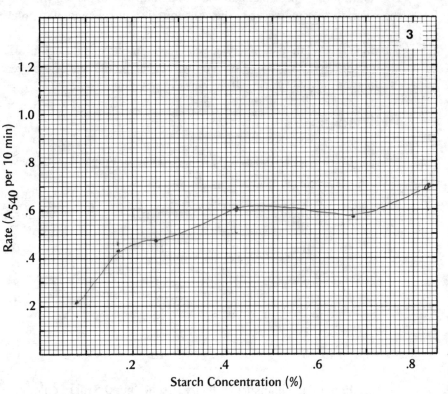

2. Did you have any concentrations of substrate that were high enough to allow amylase to work at its maximum rate? Which concentration(s)? Explain.

3. Is the rate of the enzyme-catalyzed reaction always proportional to the initial concentration of substrate?

No

IV. THE EFFECT OF ENZYME CONCENTRATION ON THE REACTION RATE

How will the concentration of an enzyme affect the rate of a reaction if the enzyme molecules can work at their maximum rate? Again there must be an abundance of substrate at the beginning of the reaction (few enzyme molecules relative to many substrate molecules). When this substrate requirement is met, the rate of reaction is proportional to the enzyme concentration. Experimentally determine the relationship between the reaction rate and the enzyme concentration.

Procedure

1. Mark three tubes 1/4E, 1/2E, and 3/4E, respectively. Then prepare varying concentrations of enzyme by making the following additions to the tubes.

Tube	Drops of NaCl Solution	Drops of Enzyme Solution
1/4E	6	2
1/2E	4	4
3/4E	2	6

   Mix the contents of each tube thoroughly. Note: Many enzymes require certain substances to be present in solution before they will work. Chloride ions must be present in order for amylase to work. In the preceding exercises, sodium chloride was already added to the enzyme solution.

2. Mark separate tubes B, 1/4, 1/2, 3/4 and 1. Add ten drops of 1% starch (pH 7) to each tube.
3. Prepare the blank (tube B) by first adding 0.5 ml color reagent and then adding 2 drops of undiluted enzyme. Mix well.
4. Initiate the reactions by adding in the following manner the different enzyme solutions prepared in step 1 to the tubes prepared in step 2:

Tube	Addition
1/4	2 drops of 1/4E
1/2	2 drops of 1/2E
3/4	2 drops of 3/4E
1	2 drops of undiluted enzyme

   Mix the contents of each tube thoroughly.
5. Allow the reactions to proceed for ten minutes at room temperature.
6. Stop the reactions by adding 0.5 ml of color reagent to each tube. Mix. Heat the tubes (including the blank) for five minutes in a boiling water bath. Add 2.5 ml of distilled water. Mix until the color is uniform.
7. Record the absorbance for each sample in Table 4-4.
8. In Graph 4, plot the absorbance values generated during the ten minute reaction as a function of the relative amount of enzyme present.

TABLE 4-4		
Tube	Relative Amylase Conc.	A_{540}
1/4	1/4	.62
1/2	1/2	.66
3/4	3/4	.66
1	1	.62

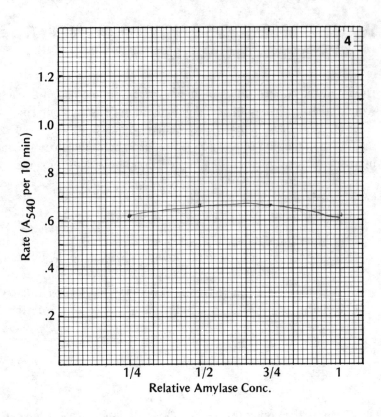

Questions

1. What is the relationship between the enzyme concentration and the reaction rate?

2. If you wanted to compare the amounts of an enzyme present in various samples of tissue, what substrate concentration should you use? Why?

3. How could you compare experimentally the relative amounts of amylase secreted by various individual organisms of the same species?

V. THE EFFECT OF pH ON ENZYMATIC ACTIVITY

You have investigated how two aspects of the chemical environment can influence the rate of an enzyme-catalyzed reaction, but what about other features of the physical and chemical environment (temperature and pH)? Enzymes are proteins, and their normal three-dimensional configuration is essential for correct enzymatic activity. Hydrogen bonds are primarily responsible for holding an enzyme molecule in its specific shape. Individually the hydrogen bonds are very weak but collectively they are strong enough to provide stability to protein structure. Therefore, anything that can disrupt these relatively weak bonds will have an effect on enzyme activity.

One aspect of the environment that can profoundly effect enzyme activity is the concentration of hydrogen ions. A measure of the hydrogen ions in solution is commonly expressed as a pH value. A pH value is the negative log of the hydrogen ion concentration. Therefore, low pH values indicate a large concentration of ions, whereas high pH values indicate a low concentration of ions. These values can range from 0 to 14. A solution with a pH below 7 is considered acidic; a solution with a pH of 7 is neutral; and a solution with a pH above 7 is basic.

The hydrogen ion concentration of a solution can be drastically altered if substances that readily ionize are added to the solution, that is, if substances are added that donate or remove hydrogen ions from solution. Those substances that contribute hydrogen ions to solution are called **acids**, and those that remove hydrogen ions are called **bases**. Therefore, in any experiment of enzyme activity it is extremely important that the experimenter be able to control the hydrogen ion concentration so it does not become an unknown variable. In the previous exercises you maintained a constant pH value in your solution although you may have been adding substances that acted either as acids or bases. To maintain a constant pH you had to add substances to your solutions called **buffers**. Buffers are substances that extract excess hydrogen ions from solution or donate ions to solution when the ion concentration is low. Solutions may be buffered to any pH. For example, if you would like to investigate the activity of an enzyme at a pH of 10, but you will be adding substances that will donate hydrogen ions to solution, buffering the solution at pH 10 will automatically remove excess hydrogen ions and maintain the specified pH.

In this exercise you will investigate the effect of the hydrogen ion concentration (pH) on the enzymatic activity of amylase. At high and low pH values many hydrogen bonds are broken in the enzyme molecule. This results in a change in the enzyme's three-dimensional structure and therefore its activity. You should note that not all types of enzyme molecules are affected equally by pH.

Why must you run a separate tube for each pH?

Procedure

1. Mark 10 tubes 3, 3B, 5, 5B, 7, 7B, 9, 9B, 11 and 11B respectively.
2. To each tube add ten drops of the appropriate starch solution (pH 3, 5, 7, 9, or 11).
3. Add 0.5 ml of color reagent and then add two drops of enzyme to each blank tube. Mix well.
4. Initiate the reaction in the experimental tubes by adding two drops of enzyme solution to each. Immediately mix the contents of each tube thoroughly.
5. Allow the reactions to proceed for ten minutes at room temperature. Then stop the reactions by adding 0.5 ml of color reagent. Mix.
6. Heat the tubes in a boiling water bath for five minutes. Add 2.5 ml of distilled water. Mix until the color is uniform.
7. Record the absorbance of each sample in Table 4-5.
8. In Graph 5, plot the absorbance values as a function of pH for each of the reaction mixtures.

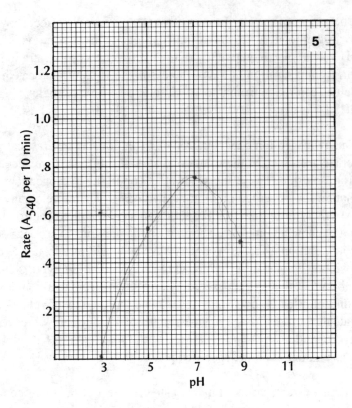

TABLE 4-5		
Tube	pH	A₅₄₀
3	3	.61
5	5	.34
7	7	.75
9	9	.48
11	11	∅

Questions

1. At which pH does amylase reach its maximum conversion rate?

   ph 7

2. Do you think that the optimum pH in a test tube is necessarily the optimal pH for maximum efficiency in a cell's metabolism? (Is the fastest rate always best?)

3. Suppose that two species contained enzymes for the digestion of starch, but that the test tube pH "optimums" were different. What would this suggest about the two enzymes?

VI. THE EFFECT OF TEMPERATURE ON THE REACTION RATE

Temperature also affects the rate of enzyme-catalyzed reactions. High temperatures cause hydrogen bonds to break, thus changing the shape of the enzyme molecule. When the shape has been disrupted, the enzyme can no longer form a functional enzyme-substrate complex. In this exercise, you will measure the reaction rate when enzyme and substrate concentrations are held constant but the temperature of the reaction is varied.

Procedure

1. Design your own experiment to test the relationship between rate and temperature. Use techniques from the previous experiments in setting up your procedure.
2. Run a reaction at each of the following temperatures: approximately 0°, 20°, 40°, 70°, and 100° C. You can use an ice bath and a boiling water bath for the extreme temperatures.
3. Record in Table 4-6 the absorbance values of each sample.
4. Plot in Graph 6 the absorbance values as a function of temperature. 10 drops 1% starch ph 7 to each tube

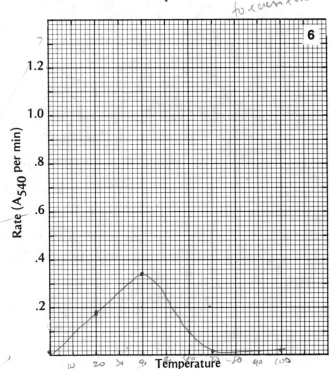

TABLE 4-6

Tube	Temp. (°C)	A_{540}
A	0	.01
B	20	.19
C	40	.36
D	70	.01
E	100	.01
B'	100	0

Questions

1. Explain why the rate of reaction is slow at low temperatures.

2. Explain why the rate of reaction is slow at high temperatures.

3. Which temperature was test tube optimal for amylase activity?

    C

4. Do you see anything significant about this optimal temperature?

    40°C

5 Respiration

INTRODUCTION

Respiration is of paramount importance to all life forms because it provides the necessary free energy to build, maintain, and perpetuate the orderly molecular arrangements of organisms. The overall process of respiration can be viewed as the sum of all the metabolic reactions concerned with the retrieval of energy stored in the chemical bonds of glucose. The energy is extracted via a series of controlled oxidation-reduction reactions, the alternating removal and addition of electrons. In biological systems, oxidation-reduction reactions often involve the removal or addition of hydrogen. Hydrogen is simply an electron in disguise, that is, an electron plus its proton. The process ultimately transforms some of the energy into a chemical form called **ATP**. ATP can then be used to drive the myriad of chemical reactions that characterize living systems.

Organisms vary in their ability to extract free energy from glucose. Some can completely oxidize glucose to carbon dioxide and water, extracting a maximum of free energy. Others are much more limited in their retrieval processes, and can extract only a small fraction of the available free energy. Those organisms that utilize oxygen to oxidize glucose to carbon dioxide and water are called **aerobes**. Those that cannot use oxygen in their respiratory processes are called **anaerobes**.

The process of cellular respiration can be thought of as occurring in four major stages; **Glycolysis, Krebs Cycle, Electron Transport System**, and the **Phosphorylation of ADP**. These stages may be distinguished, not only by WHAT they accomplish, but also by WHERE the stages occur. Note that the first stage, glycolysis, occurs in the cytoplasm of the cell, but that the remaining stages occur at specific sites within an organelle called a **mitochondrion**, Figure 5-1. I hope that knowing where these stages occur will help you organize your thoughts about the events of cellular respiration.

The extraction of energy from glucose begins with the anaerobic process of **Glycolysis**. During the first step in glycolysis, two molecules of ATP react with one molecule of glucose, Figure 5-2. This step is necessary because it provides the activation energy for the molecule to begin a journey downhill that will extract some of its free energy content. As the glucose molecule moves through the pathway, electrons are stripped off in gradual steps. The electrons are transferred to a coenzyme called NAD$^+$ (nicotinamide adenine dinucleotide), therefore reducing the coenzyme. During this reaction, two electrons and one proton are transferred to NAD$^+$, thus forming NADH. Another proton is released into the watery matrix of the cytoplasm that surrounds the glycolytic machinery. This is why NADH + H$^+$ is the correct way to describe the product of this reaction and not NADH$_2$. Note that the transfer of the electrons to NAD$^+$ is a reducing reaction, and the removal of the electrons from glucose is an oxidizing reaction.

Besides the removal of electrons, as the glucose molecule is modified into more oxidized molecules, by the breaking and making of new chemical bonds, enough energy is released to generate four molecules of ATP by a process called **substrate phosphorylation**. The ATP produced in this manner is generated by reacting ADP with high-energy sugar phosphate molecules. At the end of glycolysis, the pathway has retrieved the two ATPs initially invested, and it has produced a net gain of two ATPs. A small fraction of the total free energy of the fuel molecule is transferred to ADP and NAD+ during glycolysis. The end-product of glycolysis, two three-carbon molecules called pyruvic acid, still possesses a large amount of free energy. Therefore, pyruvic acid does not represent the lowest oxidized state resulting from the breakdown of glucose.

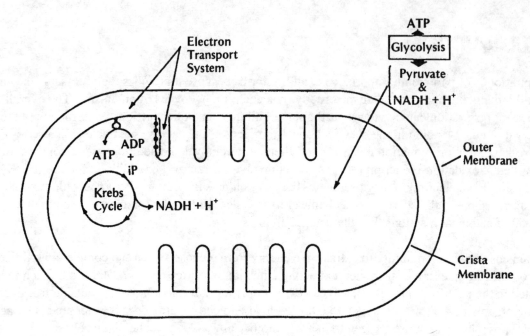

FIGURE 5-1. Mitochondrion

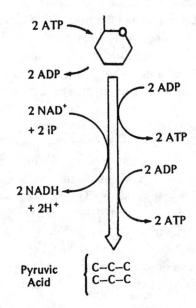

FIGURE 5-2. Glycolysis

It is at this juncture, the end of glycolysis, that aerobic and anaerobic respiration begin to differ, Figure 5-3. If oxygen is absent, some cells such as yeast convert pyruvic acid into ethyl alcohol by a process called **alcoholic fermentation**. The first step is the removal of carbon dioxide (decarboxylation) from pyruvic acid, and the production of a two-carbon intermediate compound called acetaldehyde. The electrons, in the form of hydrogens, from NADH + H$^+$ (produced during glycolysis) are then used to reduce the intermediate compound, thus converting the compound into ethyl alcohol. Other cells, such as bacteria and muscle, when deprived of oxygen, pass the electrons in the form of hydrogens from NADH + H$^+$ to pyruvic acid. The process produces lactic acid, and is called **lactic acid fermentation**. In both alcoholic and lactic acid fermentation, the only source of ATP is from glycolysis, and the net production is limited to two ATPs.

If oxygen is present in cells that possess mitochondria, the product of glycolysis is dealt with differently than in the fermentation process. In the aerobic respiratory pathway, each pyruvic acid molecule is completely oxidized to carbon dioxide and water, with the concomitant production of much more ATP than is yielded by glycolysis alone.

Since glycolysis occurs in the cytoplasm of the cell, the net ATP production can be used immediately as an energy source to drive other metabolic reactions. But, before pyruvic acid can be further oxidized, it must be moved from the cytoplasm to the site of its utilization in the interior of the mitochondrion. Once inside the mitochondrion, the pyruvic acid is stripped of two electrons (plus their

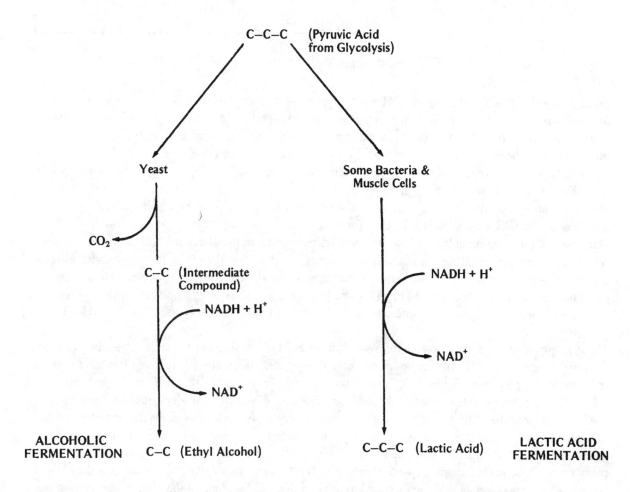

FIGURE 5-3. Alcoholic and Lactic Acid Fermentation Pathways

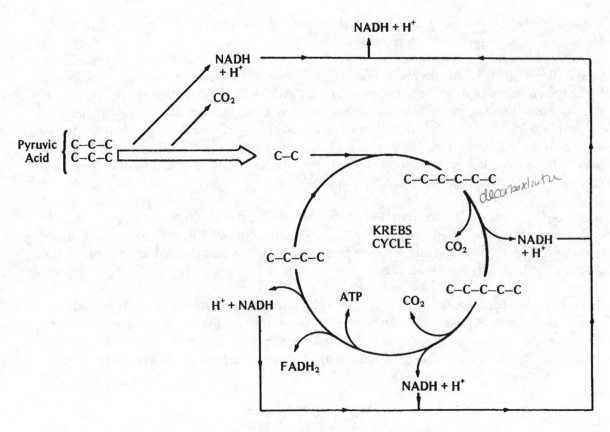

FIGURE 5-4. Krebs Cycle (Citric Acid Cycle)

protons) by NAD⁺, thus forming NADH + H⁺, Figure 5-4. During this process a molecule of carbon dioxide is released from pyruvic acid leaving behind a two-carbon molecule. This two-carbon compound then enters the **Krebs Cycle** by attaching to an existing four-compound molecule present inside the mitochondrion. The resultant six-carbon molecule next enters a series of oxidation/reduction reactions. These reactions systematically remove electrons and their protons, and transfer them to the substance NAD⁺, thus forming three molecules of NADH + H⁺. At one location in the cycle, two electrons and two protons are transferred to a coenzyme called FAD (flavin adenine dinucleotide), which is structurally similar to NAD⁺, thus forming FADH$_2$. At two other locations along their pathway, specific chemical bonds holding some carbon atoms to the chain are broken, and carbon is released in the form of carbon dioxide. This occurs at two places in the cycle, therefore shortening the original six-carbon molecule to the same four-carbon molecule that started the cycle. Also, at one point in the cycle, a molecule of ATP is produced by substrate phosphorylation. Thus, the main function of the Krebs Cycle is to produce large amounts of the high-energy molecule NADH + H⁺.

The next phase of the aerobic pathway involves a series of electron acceptor molecules embedded in the crista membrane of the mitochondrion. Collectively these molecules are referred to as the **Electron Transport System**, Figure 5-5. Each of these molecules has a different affinity for electrons, and their arrangement in the crista membrane determines the path that the electrons will follow. The NADH + H⁺ from the Krebs Cycle is present in the watery matrix adjacent to the inner surface of the crista membrane. An electron acceptor molecule, embedded in the inner surface of the crista membrane, strips off two electrons and one proton from NADH. The acceptor molecule simultaneously picks up one proton (H⁺) from the watery medium. Next, the acceptor molecule transports the electrons, and their protons, across the crista membrane to the outer surface of the membrane. Here the acceptor molecule releases the two protons into the intermembrane space formed between the crista

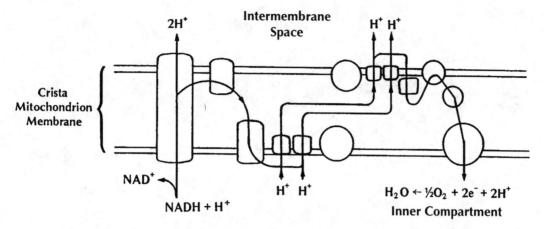

FIGURE 5-5. Electron Transport System

and outer membranes. Simultaneously, the electrons are transferred to an adjacent acceptor molecule that has a greater affinity for the electrons than the first acceptor in the chain. The electrons are again transferred, back to the inner surface of the crista membrane where they pick up additional hydrogen ions (protons).

Instead of describing all the separate electron transfers that take place in the membrane, the electrons flow through the membrane, traveling from one surface to the other via the acceptor molecules. As the electrons travel, they pick up protons from the inner compartment of the mitochondrion and then release the protons to the medium of the intermembrane space. The energy of the electrons is used to pump hydrogen ions across the membrane from the inner area of the mitochondrion to the area between the inner and outer membranes. The result of this pumping activity is the creation of a hydrogen ion gradient, with many protons outside the inner membrane and fewer on the inside. This large concentration of protons in the intermembrane space represents a reservoir of free energy, energy that can be used to power the next stage of respiration, i.e., the phosphorylation of ADP to produce ATP. Without the compartmentalization produced by the arrangement of the mitochondrial membranes, the establishment and exploitation of a hydrogen ion gradient would not be possible.

What is the fate of the electrons that pass through the Electron Transport System? The last acceptor molecule of the transport system is at the inner surface of the crista membrane. When two electrons arrive at this molecule, they must be transferred somewhere or the entire aerobic pathway, back to pyruvic acid, would come to an abrupt halt. It is here that the two electrons are passed to an atom of oxygen. The oxygen then picks up two protons from the surrounding water thus forming a molecule of water. Oxygen therefore acts as the final acceptor of the electrons stripped from glucose. This is why you, and all other aerobic organisms, need oxygen. When the electrons, from glucose, finally arrive at oxygen their energy state has been greatly reduced. It is the free energy released by the electrons, as they fall from their high energy state in glucose to their low energy state in water, that is used to synthesize ATP, Figure 5-6.

The final stage of respiration involves the actual synthesis of ATP, in a process called the **Phosphorylation of ADP**, Figure 5-7. Embedded in the same membrane as the electron transport system is the biochemical apparatus that allows the phosphorylation of ADP. Globular structures projecting into the interior of the mitochondrion house the enzymes and associated molecules that bring about the production of ATP. As the ions pass through the membrane via a protein channel, to restore equilibrium, they liberate enough energy from the proton gradient to drive the reaction leading to the formation of ATP.

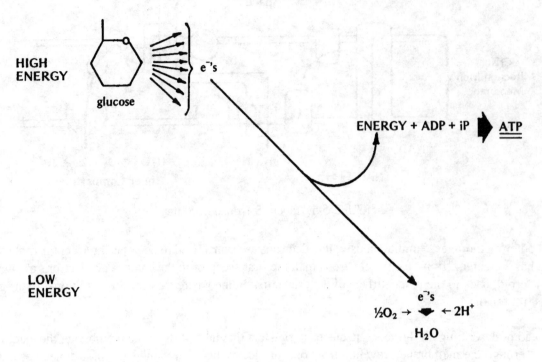

FIGURE 5-6. High Energy Electrons from Glucose Are Used to Synthesize ATP

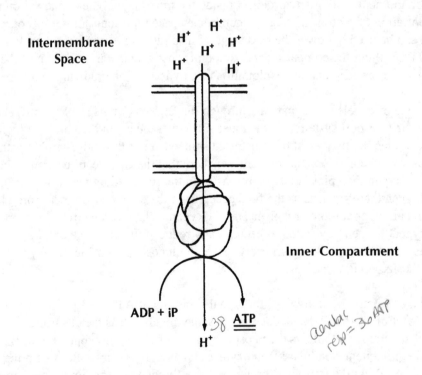

FIGURE 5-7. Phosphorylation of ADP

We have finally reached the end of the aerobic pathway, with the overall task of the pathway complete: the removal of free energy from glucose and its transfer to the terminal phosphate bond of ATP. This large supply of ATP is now readily available to supply free energy to the myriad of energy requiring reactions that characterize living forms.

Figure 5-8 illustrates how all of the separate stages of aerobic respiration fit together into a continuous pathway, which begins with a molecule of glucose and ends with glucose being oxidized to carbon dioxide and water. Do not lose sight of the fact that the overall purpose of the pathway is the production of large amounts of ATP.

The total ATP production from aerobic respiration is summarized in Table 5-1. Note: There are two major ways in which ATP may be generated; substrate and oxidative phosphorylation. The *gross* production of ATP, from these two methods, is thirty-eight molecules. To obtain the *net* yield, the two ATPs initially invested must be subtracted. This leaves a net of thirty-six ATPs. If you compare the net ATP production of aerobic respiration with that of anaerobic respiration, you will note that the former produces eighteen times as much ATP.

In the exercises that follow on aerobic and anaerobic respiration, you will not be directly concerned with the details of the metabolic processes of respiration. Rather, you will concentrate on how one environmental variable (temperature) influences the rate at which respiration occurs in three different organisms.

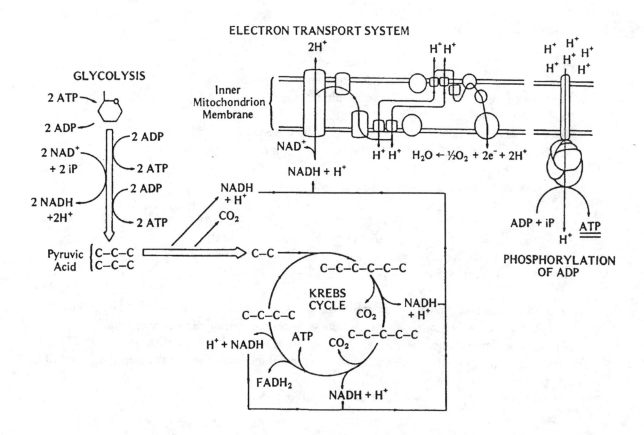

FIGURE 5-8. The Aerobic Pathway

Exercise 5 — Respiration

TABLE 5-1. ATP Production from Aerobic Respiration per Molecule of Glucose.

Substrate Phosphorylation:

Glycolysis		4 ATP
Krebs Cycle		2 ATP
	Total	6 ATP

Oxidative Phosphorylation:

Glycolysis (2 NADH + H$^+$)		4 ATP
Krebs Cycle: (8 NADH + H$^+$)		24 ATP
(2FADH$_2$)		4 ATP
	Total	32 ATP
	Gross Total	38 ATP
	Minus	2 ATP
	Net Total	36 ATP

OBJECTIVES

1. Define the following terms:

oxidation	aerobic respiration
reduction	electron transport system
glycolysis	poikilotherm
alcoholic fermentation	homeotherm
lactic acid fermentation	temperature homeostasis
anaerobic respiration	thermal neutral zone

2. Explain the role of the following in respiration:

glucose	NAD$^+$	electrons
pyruvic acid	NADH + H$^+$	ATP
mitochondrion	oxygen	Krebs Cycle
electron transport system		

3. Compare the effectiveness of anaerobic and aerobic respiration in terms of ATP production.
4. Explain graphically how different ambient temperatures would affect the respiration rate of a hypothetical poikilothermic animal.
5. Explain graphically how different ambient temperatures would affect the respiration rate of a hypothetical homeothermic animal.
6. Explain the major function of respiration in all organisms.

MATERIALS

crushed ice
CO_2 absorbant (in coin bags)
H_2O absorbant (in coin bags)
mice (in holding cage)
Tenebrio beetle larvae (meal worms)
respirometer (custom designed)
fermentation tubes of incubated yeast culture
pinch clamps

hot tap water (approx. 45°C)
1000 ml beakers
reservoir thermometers
restrainer cages
germinating bean seeds
Brodie manometer fluid
metric rule

I. ANAEROBIC RESPIRATION IN YEAST

Although the net energy yield from a glucose molecule during glycolysis is much less than the yield from glucose during aerobic respiration, many organisms use the ATP production from glycolysis as their sole source of energy. Some organisms and certain tissues in multicellular animals can carry out both aerobic and anaerobic respiration depending upon whether oxygen is available or lacking. Being able to switch between aerobic and anaerobic respiration can be useful under certain circumstances. For example, if an organism that respired aerobically suddenly found itself deprived of oxygen, it would be advantageous to be able to switch to an alternate anaerobic pathway.

Yeasts provide us with just such an example. These unicellular organisms possess all the metabolic machinery (mitochondria) to carry out aerobic respiration. When placed in an environment without oxygen, they can continue respiration via the alcoholic fermentation pathway, and meet their energy needs by producing ATP from glycolysis.

Yeasts are naturally found growing on the surfaces of fruits such as grapes, apples, etc. Can you create a possible scenario that would help explain why it would be adaptive for yeast to switch between the two types of respiration?

> **HINT**: What is the fate of most fruit that escapes being eaten? Explain below.

In the exercise that follows, you will investigate how temperature affects the rate of anaerobic respiration of yeast. Before lab, identical solutions of yeast and glucose were incubated at different temperatures in fermentation tubes. Incubation in the tubes produces anaerobic conditions. Remember that during alcoholic fermentation the yeast cells produce carbon dioxide and ethyl alcohol. The carbon dioxide is released as a gas. It can be collected and used as an indicator of the relative amounts of respiratory activity that have occurred at the different temperatures.

Procedure

On demonstration are the results of the incubation experiment. Measure the distance displaced by the CO_2 in each tube and record these distances in Table 5-2. Plot these displacements as a function of temperature on the graph provided below. Since the distances displaced in the tubes are directly proportional to the volumes of CO_2 produced, it is not necessary to calculate the actual volumes for a comparison of fermentation activity. The distances are sufficient.

TABLE 5-2. Carbon Dioxide Production During Fermentation.

Temp. °C	Dist. Displaced in Column (cm)
10°	.2
20°	2.6
30°	6.7
40°	8.8
50°	6.0
60°	1.2
70°	.3

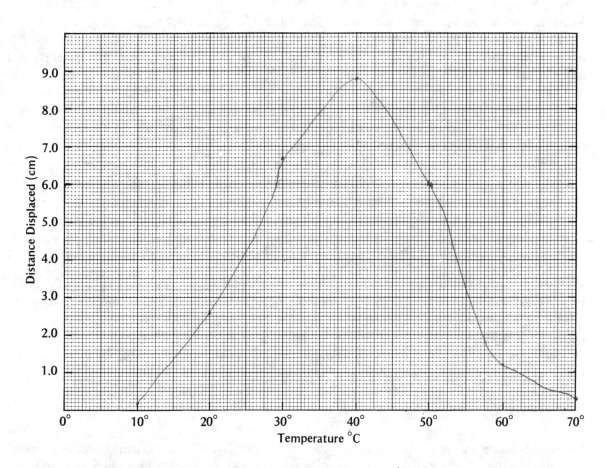

Plot of Distance Displaced by CO_2 as a Function of Temperature

Questions

1. Do your graphed results show an approximate exponential increase in CO_2 production with an increase in temperature up to 40°C? _____
   Why?

2. Speculate about why the maximum rate of alcoholic fermentation occurred near 40°C. Your instructor may need to help you with this one!

3. Can you think of any reason(s) why glycolytic activity tapers off rapidly at temperatures above 40°C, and is near zero at 60°C?

4. If glycolysis does not take place at 70°C, how would you explain the presence of a small amount of CO_2 in the 70°C fermentation tube?

II. AEROBIC RESPIRATION

Concurrent with the evolution of larger and more mobile organisms were the evolution of metabolic pathways that could extract large amounts of energy, in the form of ATP, from glucose. Remember, the glycolytic pathway is limited to the production of only two net ATPs per molecule of glucose, and that pyruvic acid is a long way from being completely oxidized to carbon dioxide and water. Therefore, any metabolic pathway that could carry the oxidation of glucose a step further, with the concomitant production of more ATP, would have provided an organism bearing this type of pathway with the necessary energy to support new physiological, structural, and behavioral adaptations to its environment. The metabolic pathway that makes up aerobic respiration is just such a pathway. Using oxygen to completely oxidize glucose to carbon dioxide and water, aerobic respiration can potentially generate 18 times more ATP than anaerobic respiration.

Though aerobic respiration is much more effective in generating ATP than anaerobic respiration, approximately 60% of the energy released during the oxidation of glucose to carbon dioxide and water escapes capture as chemical bond energy, and is given off as heat. Because most organisms have high rates of thermal conductance, most of this heat is lost to the surroundings as rapidly as it is produced. Those organisms that rapidly lose this heat energy, and therefore have their body temperatures regulated primarily by the external environment, are called **ectotherms**. The heat that determines their body temperature is acquired from the environment, and not from their own metabolic activities. Because the external environment largely regulates the temperature of ectotherms, their body temperatures can be highly variable, depending upon the ambient temperature. Organisms that have a variable body temperature are referred to as **poikilotherms**. The activity of ectotherms is greatly affected by changes in the external environmental temperature. Increased external temperature usually allows a corresponding increase in activity, whereas a depression in temperature usually leads to a decrease in activity.

In contrast to the ectotherms, which includes the vast majority of organisms, a few groups of organisms can regulate body temperature using their metabolic heat energy. These organisms have mechanisms that lower the thermal conductance of their body surfaces, so that heat loss is retarded to the external environment. These organisms are called **endotherms**. Those endotherms that can **continuously maintain** an **elevated** and **constant** internal temperature are called **homeotherms**. An elevated and constant internal temperature permits a constant, high-level of both metabolic and physical activity that is somewhat independent of external temperature. Homeothermic animals are therefore able to exploit and be active in habitats that are usually not available to poikilotherms. Homeotherms include the birds and mammals, but not all are continuously endothermic. Can you think of a few examples? The ability to be partially endothermic is also seen in some large reptiles and very large fast swimming fishes. Even some insects, such as honeybees, can maintain body temperatures that are higher than the ambient temperature during periods of activity.

To help produce and maintain a stabilized and elevated internal temperature, homeotherms have evolved many **structural**, **physiological** and **behavioral** mechanisms. For example, birds use feathers and mammals use hair to help retard the exchange of heat with their surroundings, thereby insulating themselves from the external environment. They have also evolved the capacity to increase the output of metabolic heat when these mechanisms have reached their limit in preventing heat loss. Shivering, which is a high rate of muscular activity driven by ATP, is an example of a heat-generating adaptation. Because large amounts of ATP are consumed by the muscular contractions during shivering, an increased demand for metabolic energy is placed on the aerobic pathway. Since heat is a by-product of metabolism, an increase in the metabolic rate produces heat that can be used to maintain body temperature.

Beyond the structural and physiological mechanisms, the behavior of an organism can contribute greatly to the regulation of body temperature. For example, when a dog is cold, it minimizes its surface to volume ratio by curling up. This effectively reduces the surface area from which heat can be lost to the surrounding environment. Conversely, the animal sprawls out on hot days which maximizes the surface area from which heat can dissipate.

In the exercise that follows, you will investigate the effect of different external temperatures on the respiration rates of a small homeotherm, the house mouse, and two ectotherms, a plant (germinating bean seeds) and an invertebrate (meal worm beetle larvae). You should also observe how the structural and behavioral adaptations of the mouse help it regulate its internal temperature.

General Instructions

You will use a **respirometer** to measure changes in the respiration rate of your organisms. The respirometer is designed to measure the amount of oxygen consumed per unit time under various ambient (surrounding) temperatures. You can use oxygen consumption as an index of respiratory activity because more oxygen is consumed as respiratory activity increases.

The respirometer you will use consists of two major components: the respiratory chamber and the manometer tube, Figure 5-9. A restraining cage inside the chamber supports a mouse above substances that absorb CO_2 and H_2O, the two by-products of respiration. As oxygen is consumed by the organism, and the CO_2 and H_2O are absorbed, the gas pressure of the atmosphere inside the chamber is reduced. If you can measure this decrease in gas pressure, you have measured the amount of oxygen consumed in respiration.

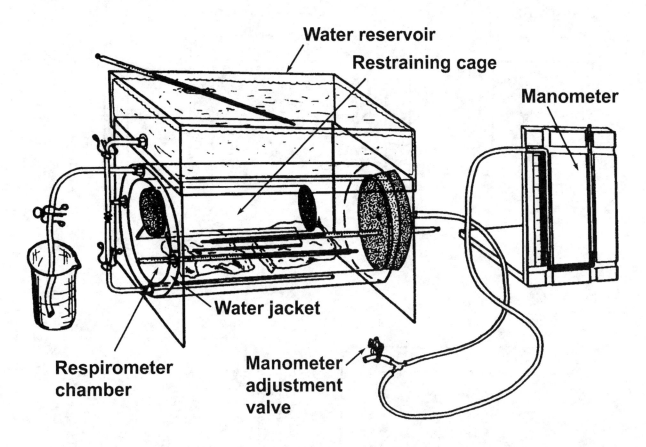

FIGURE 5-9. Respiration Apparatus

To detect this change in pressure, an U-shaped tube (manometer) filled with a special fluid is attached to the respiratory chamber. When the pressures exerted on the two ends of the column of fluid are equal, the levels of the liquid will be equal, Figure 5-10a. If the pressure inside the chamber decreases, as occurs when oxygen is consumed by an organism, the fluid will move toward the chamber, Figure 5-10b. An increase in pressure inside the chamber occurs when air inside the chamber is heated. This will cause the fluid to move away from the chamber, Figure 5-10c. A valve between the chamber and manometer tube allows you to initiate a reading, and to reestablish equal fluid levels without removing the end of the chamber, when the chamber has reached temperature equilibrium.

Exercise 5 — Respiration

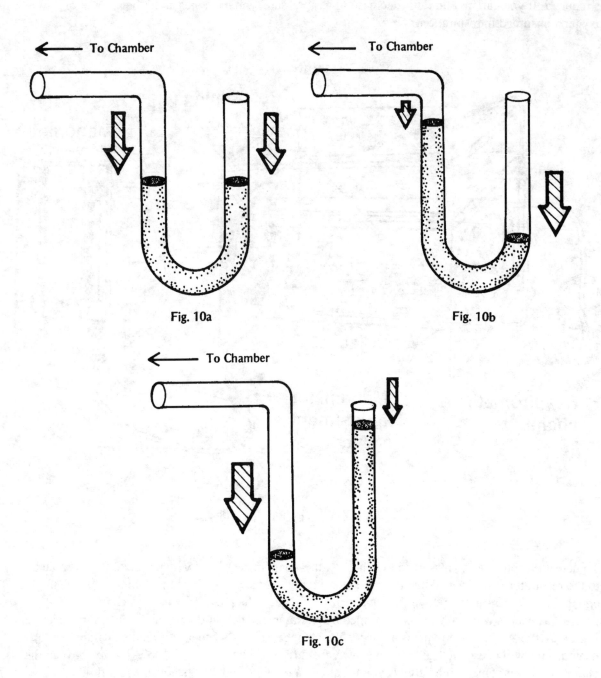

FIGURE 5-10. Manometer Tubes

A waterjacket around the respirometer is used to regulate the temperature inside the chamber. It is fed by a reservoir that sits atop the respiratory chamber, and should be filled with water at the appropriate temperature. The success of this experiment depends largely on how well you can manipulate and maintain an experimental temperature inside the chamber. Therefore, it is important that you follow the directions closely, and that you listen to your instructor for helpful suggestions.

The most common mistake made at the warmest temperature is a failure to halt heat flow into the respiratory chamber. Failing to halt this heat flow causes the gas in the chamber to expand. This heat creates more pressure, counteracting the decrease in pressure caused by the consumption of oxygen by the mouse. Therefore, there is a net increase, rather than decrease in pressure. This is evident by the movement of the manometer fluid in a direction away from the respiratory chamber. Therefore you must be certain that you have stopped the heat flow into the chamber.

NOTE: You should be aware that heat is being exchanged between three types of media (gas, liquid, and solid), and that the rate of heat exchange at each of the interfaces of these media is different. Therefore, you must adjust the temperature in the waterjacket with the correct amount of lead or lag time to compensate for the different rates of heat exchange.

Each respirometer should have three thermometers; one for the water reservoir, another inserted permanently in the waterjacket housing, and a third inserted in the chamber stopper. You should seal the chamber so the chamber thermometer can be read simultaneously with the waterjacket thermometer. Use care in inserting the stopper. You can easily break the attached thermometer.

The slightest leak in the chamber will cause misleading results or a totally inoperable system. Therefore, it is extremely important that the chamber be sealed securely. To obtain an airtight seal, place the stopper in the end of the chamber with the thermometers lined up roughly adjacent to each other. Press firmly and uniformly on the stopper. There is no need to exert great force. Lightly twist the stopper once it is pressed firmly into place. This will have the effect of locking the stopper into place.

> **CAUTION**: Whenever you stopper the chamber, THE PINCH CLAMP ON THE MANOMETER LINE MUST BE OPEN! If you fail to open the pinch clamp, the manometer fluid will be blown out of the glass U-tube.

By keeping the reservoir nearly full of water, you will ensure that you have the maximum rate of flow into the waterjacket. If you have problems getting water to flow from the reservoir to the waterjacket, simply squeeze the hose between the waterjacket and reservoir. This should eliminate any air pockets that have accumulated in the hose. To remove the reservoir, clamp the waterjacket/reservoir hose above and below the plastic connector, and then separate the connecting junction. Do not remove the hose from the attachment points on the chamber or reservoir. NOTE: MAKE NO ALTERATIONS IN THE EQUIPMENT. The thermometers, hose connections, etc., are all properly aligned and fastened. If you suspect that your equipment is faulty, consult your instructor. At the end of the experiment, empty the respirometer of water, clean the apparatus, including the restraining cage. Please leave the apparatus in the proper condition for the next group of students.

The class will be divided into groups of four and assigned an experimental temperature. Each group will be responsible for measuring the respiration rate of a mouse, germinating bean seeds, and meal worms at their experimental temperature. Oxygen consumption values for the other experimental temperatures should be shared among the groups. Be sure that your group obtains the results from the other groups before you leave lab.

Assembling the Apparatus: Figure 5-9

1. Place the restraining cage, bag of carbon dioxide absorbant, and bag of water absorbant in the respiratory chamber.

2. Insert the large rubber stopper into the end of the chamber with the thermometers lined up roughly adjacent to each other.

 CAUTION: Be certain the vent tube is open whenever you stopper the chamber. If you fail to have the vent tube open, the fluid will be blown out the glass U-tube.

Now, firmly and uniformly press on the stopper. There is no need to exert great force. Slightly twist the stopper while exerting moderate pressure once the stopper has been seated into place.

3. Make certain that the tubing from the manometer is connected to the plastic t-connector.

4. Connect the reservoir outlet to the jacket inlet.

5. Position the overflow container to catch the water from the waterjacket exhaust.

Filling the System and Achieving Temperature Equilibrium:

The general procedure is to initially fill the reservoir and waterjacket with water that is MUCH WARMER than the experimental chamber temperature of 30°C, or MUCH COOLER than the experimental chamber temperature of 15°C. These large differences in temperature allow for rapid heat exchange between the waterjacket and the respiratory chamber and should minimize the time necessary to reach the experimental temperature. Once your respiratory chamber is near the experimental temperature, it will be necessary for you to adjust the reservoir temperature to the experimental temperature. For 15°C, you will need to add warm tap water, and for the 30°C you will need to add either cold tap water or ice. For the meal worms and germinating bean seeds at the experimental temperature of 22°C (room temperature), you should use tap water close to 22°C. For the mouse at 22°C, you should use slightly cooler tap water (approximately 20°C). Why would you want to use slightly cooler water in the waterjacket when measuring the respiration rate of the mouse?

Remember that it takes 5-10 minutes to totally replace the volume of water in the waterjacket, so you must anticipate the necessary lead and lag time required. For example, if the waterjacket temperature is 2-3°C cooler than the experimental temperature of 15°C, and you initiate your reading, heat will still be flowing from the chamber to the waterjacket. The pressure in the respiratory chamber will therefore be decreased. In this situation your readings will be less than they should be or perhaps even negative. Therefore, you must ensure that the waterjacket and chamber are at temperature equilibrium.

Procedure

1. Clamp the tube leading from the reservoir to the waterjacket.

2. Fill the reservoir with water at the appropriate temperature:

Experimental Temp.	Initial Reservoir Temp.	Approx. Time to Equil.
15°C	3-7°C	30 min.
22°C	22°C	1-2 min.
30°C	40-45°C	30 min.

For 15°C, combine ice and tap water to make a slurry in the reservoir. Keep ice away from the reservoir outlet to maintain a constant flow of water into the waterjacket. Replenish the reservoir with slurry as needed. At 30°C, fill the reservoir using hot tap water (40-45°C). If the water from the tap is not hot enough, have your TA warm water on a hot plate. Never use water hotter than 60°C in the reservoir. For 22°C, use tap water adjusted to this temperature.

3. Release the pinch clamp between the reservoir and waterjacket, and squeeze the tube between your fingers a few times to evacuate the line of air. Be certain that the waterjacket fills completely and that the overflow container is in place.

4. As the waterjacket fills, replenish the reservoir with water of the prescribed temperature.

6. Monitor the waterjacket and respiratory chamber temperatures. As you approach your experimental temperature, within 2-3°C, begin to adjust your reservoir temperature accordingly. Again, remember that it takes several minutes to replace the water in the waterjacket, so allow for adequate time.

7. When you think that the respiratory chamber and waterjacket are at the same temperature, check for temperature equilibrium by simply pinching shut the venting tube with a clamp. Temperature equilibrium has been achieved when there is no change in the manometer fluid level. If the manometer fluid moves away from the respiratory chamber, heat is still flowing into the chamber causing an increase in pressure. In this case, the waterjacket temperature is too high. If fluid movement is toward the chamber, the waterjacket temperature is too cold and heat is flowing out of the respiratory chamber, thus decreasing the chamber pressure. You should make the necessary adjustments to correct these temperature differences.

Important !!
At this point check to see that:
   a. The waterjacket and chamber temperatures are at temperature equilibrium.
   b. The reservoir is full of water at the appropriate temperature.
   c. The absorbant bags and restraining cage are inside the chamber.

If these things are all in place, you are ready to insert your organisms.

Loading the Mouse: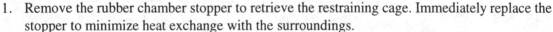

1. Remove the rubber chamber stopper to retrieve the restraining cage. Immediately replace the stopper to minimize heat exchange with the surroundings.

2. Load the mouse into the restraining cage. Hold the mouse lightly by the tail over the cage so that its front feet touch the lower edge of the opening. Tip the cage up and the mouse will walk uphill into the cage. Stopper the end once the mouse is inside.

3. MAKE CERTAIN THAT THE VENT LINE IS OPEN AT THE T-CONNECTOR. Open the chamber and insert the restraining cage with the breather holes oriented toward the large rubber stopper. Quickly insert the chamber stopper, and align the thermometers. Firmly and uniformly press on the stopper. Once the stopper has been seated into place, give it a firm twist while applying moderate pressure. This should seal the chamber ensuring that it is airtight.

4. Allow a few minutes for the chamber and waterjacket to reestablish equilibrium. Once temperature equilibrium has been reestablished, move on to the section, "Running the Experiment."

Loading the Meal Worms:

1. Weigh one plastic bag of meal worms (it should weigh approximately 30g). Record this value in Table 5-5.

2. Remove the chamber stopper and quickly insert the bag of meal worms directly into the chamber.

3. MAKE CERTAIN THAT THE VENT LINE IS OPEN AT THE T-CONNECTOR. Quickly insert the rubber stopper, and align the thermometers. Firmly and uniformly press on the stopper. Once the stopper has been seated into place, give the stopper a firm twist while applying moderate pressure. This should seal the chamber so it is airtight.

4. Allow a few minutes for the chamber and waterjacket to reestablish equilibrium. Once temperature equilibrium has been reestablished, move on to the section, "Running the Experiment."

Loading the Germinating Bean Seeds:

1. In a plastic bag, weigh out germinating bean seeds (approximately 150 g). Enter this weight in Table 5-6.

2. Remove the chamber stopper, and remove the restraining cage, if necessary. Stopper the chamber immediately to minimize heat exchange with the surroundings.

3. MAKE CERTAIN THAT THE VENT LINE IS OPEN AT THE T-CONNECTOR. Open the chamber and insert the bag of germinating seeds. Quickly insert the chamber stopper, and align the thermometers. Firmly and uniformly press on the stopper. Once the stopper has been seated into place, give the stopper a firm twist while applying moderate pressure. This should seal the chamber so it is airtight.

4. Allow a few minutes for the chamber and waterjacket to reestablish equilibrium. Once temperature equilibrium has been reestablished, move on to the next section, "Running the Experiment."

> **SPECIAL NOTE**: As the organisms adjust to their new environment during the next few minutes, decide on the duties to be carried out by the group members (timer, recorder, organism and temperature observer, and pinch clamp releaser).

Running the Experiment:

The goal of the experiment is to measure the amount of oxygen consumed by each organism at the different experimental temperatures. Oxygen consumption is indicated by a movement of manometer fluid up the scale, toward the chamber. You will not be measuring the absolute volume of oxygen consumed per minute, but the vertical displacement of the manometer fluid as a relative indicator of the oxygen consumed.

You will need to measure the amount of oxygen consumed by the meal worms and germinating plant seeds for approximately 20-30 minutes. A single reading over the duration of this period should be sufficient. Oxygen consumption by the mouse is much greater per unit time than for the other organisms, so you will take readings over a shorter period (10 minutes). Because the mouse consumes more oxygen during 10 minutes than is contained in the length of the manometer tube adjacent to the measuring scale, you will need to take multiple readings during the 10 minute intervals. Therefore, during the run it will be necessary for you to periodically reset the manometer fluid back to its original level. You will need to keep track of these separate readings in Table 5-3, and then add them together to obtain the value for total displacement during the 10 minute periods.

During each run, you will record the baseline fluid level, height of fluid at time of each venting, displacement during each venting period, and the time of each venting. You will also record the initial and final temperatures of the chamber and waterjacket at the beginning and end of each run, respectively. Record all of the above information in Tables 5-3, 5-5, and 5-6. In Table 5-4, record your behavioral observations for the house mouse. Once you have collected all your data, you will need to calculate the amount of oxygen consumed per gram weight. This is expressed as cm of displacement per minute per gram. Share this data with other groups and obtain their data. Record all data in Table 5-7, then plot these results on the graph in Figure 5-11.

Executing the run:

1. Record the initial manometer fluid level in the appropriate Table (Table 5-3 for mouse, Table 5-5 for meal worms, or Table 5-6 for bean seeds).

2. To initiate a run, fold the vent tube over and clamp it shut. Begin timing.

    a. If there is an immediate negative flow of the manometer fluid (away from the scale), release the pinch clamp and repinch the tube. Restart the timing.

    b. If there is no flow in the positive direction (up the scale) after five minutes, release the pinch clamp. Reinsert the chamber stopper more firmly, and again clamp the t-connector. Restart the timing.

3. If the manometer fluid level reaches 6 cm, on the ruler scale, be prepared to release the pinch clamp on the t-connector. Please note that generally the fluid rises slowly in the beginning and then rapidly accelerates as it reaches the 6 cm mark of the scale.

4. Never let the manometer fluid go beyond 8-9 cm on the ruler scale. During the mouse run it will be necessary for you to take multiple readings during the 10 minute runs. Release the vent tube clamp momentarily to reset the manometer fluid, then reclamp the hose and continue the run. Record the time elapsed, and the total height that the manometer fluid reached during each interval in Table 5-3. Repeat steps 2 and 3 for the duration of the 10 minute run. For the meal worms and bean seeds, you can probably obtain results from a single continuous run (approximately 20-30 minutes).

 CAUTION: If you should fail to vent the apparatus in time to reset the manometer, and fluid is pulled into the manometer tubing leading to the t-connector, immediately clamp off the tube between the rubber stopper and the t-connector. You will need to replace the manometer and tubing. Ask your instructor for assistance.

5. Also during the mouse run, carefully observe the mouse, and note the general level of activity, and any other changes in such things as body posture, color of exposed skin, and other behavioral activities. Record these observations in Table 5-4.

6. When you have completed the experiment for each organism, release the pinch clamp on the vent tube. Remove the organisms from the respirometer. Before you return the mouse to the holding cage, weigh the mouse while in the restraining cage, then return it to the holding cage. Return to the scale and weigh the restraining cage and stopper. Subtract the total weight of the restraining cage and stopper from the weight of the mouse plus cage to obtain the weight of the mouse. Record this in Table 5-3.

7. Calculate the displacement of manometer fluid for each reading interval as follows and record these values in Table 5-3.

    Height at Venting minus Baseline Fluid Level = Displacement

8. To calculate the average displacement, per minute for each organism, divide the total observed displacement by the total time of the run. Record the results and the final waterjacket and chamber temperatures in the appropriate Table. Transfer the results of your experiment to Table 5-7. Also, obtain the experimental results from the other groups for each organism and temperatures, and record the results in Table 5-7. Use these values to plot oxygen consumption curves for each organism in Figure 5-11.

9. When you have completed all the experiments, clean the restraining cage and drain your respirometer of water. Stopper the chamber, but leave the t-connector vented.

TABLE 5-3. Student Data Collection—Mouse

Initial chamber temp. __23__ °C;   Initial waterjacket temp. __25°C2°__

Baseline Fluid Level	Height at Venting (cm)	Displacement (cm)	Time at Venting
1.8	3.0	1.2	10
1.7	6.0	4.3	17
1.7	6.0	4.3	22
1.7	6.0	4.3	27
	Total	14.1	10 min.

Final chamber temp. __23__ °C: Final waterjacket temp. __25__ °C

Mouse weight __22__ (g)

Average displacement per minute __.52__ (cm/min.)

Average displacement per minute per gram __.0237__ (cm/min./g)

TABLE 5-4. Behavioral Observations—Mouse

Suggested Chamber Temp.	Actual Chamber Temp.	Behavioral Observations
15°	16	turns paler, more behavior
22°	25	normal
30°	35	turns a brighter pink, slow movement — endo homeo

TABLE 5-5. Student Data Collection—Meal Worms excto

Initial chamber temp. __22__ °C;   Initial waterjacket temp. __24__ °C

Baseline Fluid Level	Height at Venting (cm)	Displacement (cm)	Time at Venting
1.8	3.2	1.4	20

Final chamber temp. __22__ °C: Final waterjacket temp. __24__ °C

Meal worm weight __41__ (g)

Average displacement per minute __.67__ (cm/min.)

Average displacement per minute per gram __.0017__ (cm/min./g)

Exercise 5 — Respiration

TABLE 5-6. Student Data Collection—Bean Seeds

Initial chamber temp. _____°C;   Initial waterjacket temp. _____°C

Baseline Fluid Level	Height at Venting (cm)	Displacement (cm)	Time at Venting
_____	_____	_____	_____

Final chamber temp. _____°C: Final waterjacket temp. _____°C

Bean seed weight _____(g)

Average displacement per minute _____(cm/min.)

Average displacement per minute per gram _____(cm/min./g)

TABLE 5-7. Aerobic Respiration: Summary of Results

Organism	Suggested Chamber Temp.	Actual Chamber Temp.	Average Displacement per min.	Average Experimental Displacement per min. per gram
mouse	15°			
meal worms	15°			
bean seeds	15°			
mouse	22°			
meal worms	22°			
bean seeds	22°			
mouse	30°			
meal worms	30°			
bean seeds	30°			

Discussion

Figure 5-12 is a plot of the body temperatures of a hypothetical homeotherm and a poikilotherm over a range of environmental temperatures. Notice that the poikilotherm has a body temperature that is the same as the environment, while the body temperature of the homeotherm is constant over the temperature range and is independent of environmental temperature. Therefore, the poikilotherm is expending little or no energy on thermoregulation of its body temperature, while the homeotherm must continuously expend metabolic energy. If oxygen consumption is plotted for these two hypothetical organisms, Figure 5-13, we see that the curves look very different. For the poikilotherm, the oxygen consumption curve looks like a plot of the rate of an in vitro chemical reaction. That is, for every 10°C increase in temperature, the rate doubles. Therefore, oxygen consumption for the poikilotherm appears exponential. In contrast, oxygen consumption of the homeotherm decreases with an increase in temperature up to some point, then is somewhat independent of further increases in environmental temperature. This means that at cold temperatures, the homeotherm must expend more energy to maintain its body temperature. As temperature increases, the demand for energy decreases to a point where, over a small temperature range, energy consumption for thermoregulation is at a minimum. This range of temperatures where energy consumption for thermoregulation is at a minimum is called the **thermal neutral zone**.

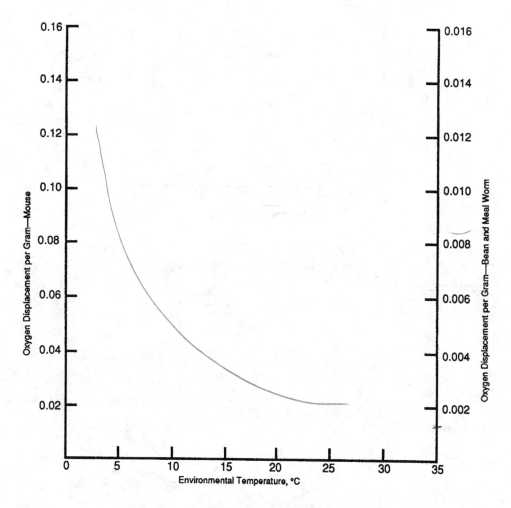

FIGURE 5-11. Oxygen Consumption of the House Mouse, Meal Worms, and Germinating Bean Seeds at Different Ambient Temperatures

Does the plot of your experiment results in Figure 5-11 look somewhat similar to the plot of oxygen consumption relative to temperature in Figure 5-13? If not, can you explain why your curve looks different?

Homeotherms can maintain a stable temperature within the thermal neutral zone, with a minimum expenditure of metabolic energy, because they can regulate body temperature with adaptations that are metabolically low cost. For example, by changing body posture, an animal can increase or decrease the surface area from which heat can be either gained or lost.

Did you notice any changes in the behavior of your mouse at the various ambient temperatures that would indicate that it was trying either to conserve or dissipate heat? Under the appropriate headings below, list some of these changes.

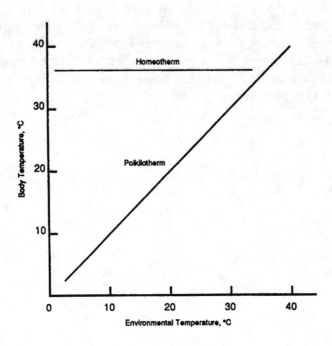

FIGURE 5-12. Body Temperature of Homeotherm and Poikilotherm in Relation to Enviromental Temperatures

Cold Temperatures:

    1.

    2.

    3.

Warm Temperatures:

    1.

    2.

    3.

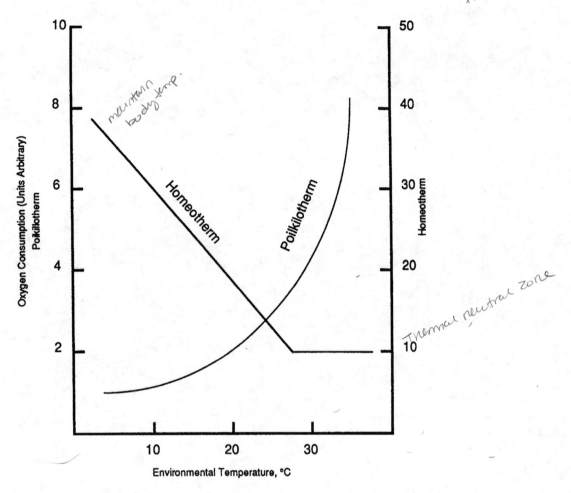

FIGURE 5-13. Oxygen Consumption of Homeotherm and Poikilotherm Relative to Environmental Temperatures

Do you see any parallels in behavior between mouse and man when both are subjected to extremes in external temperature? If you do, why would you expect to see these similarities?

Why can many reptiles, such as snakes, survive on one feeding per month while mammals of the same body mass starve in less than one week?

What is the main function of oxygen in aerobic respiration?

6 Photosynthesis

INTRODUCTION

When we compare the processes of photosynthesis and respiration, we see they are opposite sides of the same coin, Figure 6-1. That is, the respiratory pathways extract energy from organic compounds, while the photosynthetic process captures energy during the manufacture of the compounds used as fuel in respiration. Also note that the inputs to the photosynthetic process (water and carbon dioxide) are the outputs of the aerobic respiratory pathway, making these truly interconnected and interdependent global processes.

At first glance, one might develop the impression that photosynthesis was one of the first major metabolic processes to have evolved because it generates the organic compounds consumed as fuel in respiration. Contrary to this interpretation, evidence suggests the opposite was true, that is, respiration preceded photosynthesis in evolutionary history.

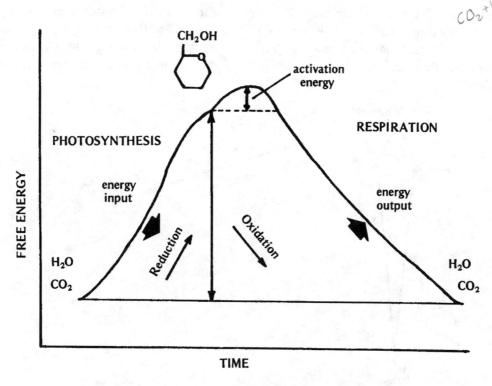

FIGURE 6-1. Photosynthesis and Respiration Compared

Exercise 6 — Photosynthesis

It is thought that perhaps some 3.8 billion years ago, the primitive oceans were filled with enormous quantities of organic compounds. These compounds had been synthesized by abiotic processes in the atmosphere and in secondary reactions in aquatic environments. The compounds made in this manner could accumulate because the reducing atmosphere produced thermodynamically stable conditions. When the first life forms arose, there would have been an abundance of food (fuel) molecules for these organisms to use. Eventually the amount of food in the environment would have become finite because the abiotic synthesis of organic compounds would have stopped due to the transition of the atmosphere to a more oxidized state. This in turn would have produced thermodynamically unfavorable conditions for the formation and stability of additional organic compounds. Competition for the finite supply of food would have increased as life forms proliferated in the oceans. In such a competitive environment, any energy procuring system that could use an alternate energy source would have had a tremendous selective advantage. The photosynthetic process is just such a system. The process uses the radiant energy of the sun to manufacture its own organic fuel compounds. Organisms no longer had to rely on a reservoir of previously manufactured organic compounds for their energy needs. What gradually emerged on the face of the planet was a collection of organisms; some that used sunlight to manufacture the organic compounds they needed (photoautotrophs), and others (heterotrophs) that satisfied their energy requirements by consuming, directly or indirectly, the organic compounds produced by the photoautotrophs.

Because the photosynthetic process is so intimately tied to all life forms, you will look in some detail at how the process functions, and how it is organized in eukaryotic plants. The process can be broken down into two main events, the **light** and the **dark reactions**. The light reactions are chiefly concerned with the initial capture of radiant energy, and its conversion and transfer to chemical bond energy, Figure 6-2. Eventually some captured energy is converted into the chemical bond energy housed in ATP and NADPH, the major products of the light reactions. The dark reactions then use the energy in ATP and NADPH to incorporate CO_2 into the existing molecular structure of a plant cell. More specifically, CO_2 is reduced to glucose. The newly formed chemical bonds in glucose represent part of the energy contained in the ATP and NADPH molecules, and therefore originally present in sunlight. The chemical bonds in a glucose molecule represent sources of energy that can be called upon to power further manipulations of matter by the plant or by an organism that would consume the plant for food.

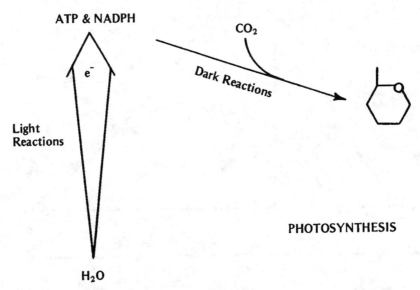

FIGURE 6-2. The Light and Dark Reactions of Photosynthesis

Before we begin a discussion of the light and dark reactions, it might be helpful if you know where the reactions occur inside a plant cell, Figure 6-3. The site of photosynthesis in eukaryotic plant cells is a membrane-bound structure called the **chloroplast**. It is separated from the cytoplasm by an outer envelope that consists of two membrane layers. In the interior of the organelle is another membrane that forms flattened sacs called **thylakoids**. The thylakoid sacs or discs are often stacked into structures called **grana**. Embedded in and on the thylakoid membranes are many of the molecules, such as chlorophyll and electron acceptor molecules, that participate in the light reactions. The orientation of these molecules embedded in the membrane plays an important role in the events of the light reactions. The specific placement of these molecules is not very pertinent to a general discussion of the light reactions. Therefore, suffice it to say the membrane holds the molecules in a specific orientation to each other. Without this correct orientation the process could not occur. The significance of this orientation should be apparent shortly.

The space between the outer envelope and the thylakoid membranes is called the **stroma**. The stroma is the site of the dark reactions. It also contains the chloroplast's ribosomes and genetic material (DNA).

The Light Reactions

The focal point of the light reactions is the movement of electrons from and through the series of molecules embedded in the internal membrane of the chloroplast. The journey of the electrons originates in a complex of chlorophyll and various other pigment molecules collectively called a **photosystem**. This complex of molecules is thought to act as an antenna that gathers in radiant energy from the sun, and then focuses the energy toward a central chlorophyll molecule. It is this central chlorophyll molecule that contributes the electrons to the light reaction pathway. The absorbed light energy is used to raise the electrons to a higher energy level.

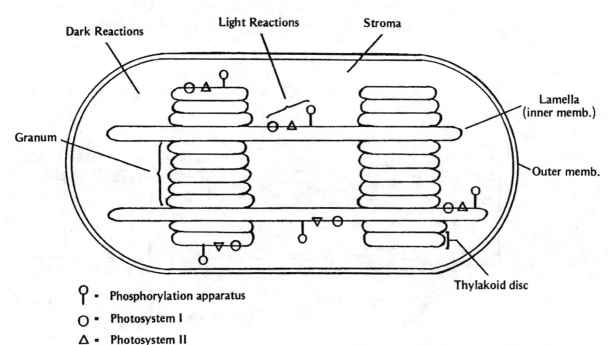

FIGURE 6-3. Structure of a Chloroplast

Once the electrons from a chlorophyll molecule have been raised to a sufficiently high energy level, an electron acceptor molecule embedded in the membrane adjacent to the chlorophyll complex grabs onto the electrons. Besides this first electron acceptor molecule, there is a whole series of different acceptors embedded in the membrane. Because of the specific placement of the different acceptor molecules adjacent to each other, and their differing affinities for electrons, a series of molecular steps is created by which the electrons are passed, successively, from one acceptor molecule to another. As the electrons move from acceptor molecule to acceptor molecule, the energy they possess decreases as the electrons occupy exceedingly lower energy levels with each transfer. Note, that as long as the movement of electrons is in the direction of lower energy levels, the process can take place spontaneously, i.e., without an external source of energy.

In the light reactions, the electrons are "excited" from their ground state in a photosystem to a higher energy level. Upon reaching this higher energy level, the first acceptor molecule in the chain grabs the electrons. Almost immediately, another acceptor molecule adjacent to the first grabs the electrons. This gaining then losing the electrons (**oxidation/reduction**) continues along the chain until the electrons finally return to their "ground state" in the chlorophyll molecule. This process is called **Cyclic Electron Flow** because the electrons return to their original location in the photosystem, Figure 6-4. As the electrons descend to their original positions, the energy they absorbed must be liberated. It is the energy liberated by the electrons as they descend through the acceptor chain that drives the production of one of the major products of the pathway—**ATP**.

The specific mechanism for the synthesis of ATP has not been completely elucidated, but a fair picture does exist. As the electrons move from adjacent acceptor to adjacent acceptor molecule embedded in the thylakoid membrane, they extract hydrogen ions from the watery medium outside the thylakoid membrane. As the electrons move from the external surface to the internal surface of the membrane, they pump the hydrogen ions to the interior of the thylakoid sacs. In doing so, a store of energy is established in the hydrogen ion gradient that results. This store of energy takes on two forms. The first form is represented by the difference in hydrogen ion concentration as measured by pH. Energy is also stored in the electric charge carried by the protons. Therefore, the difference in

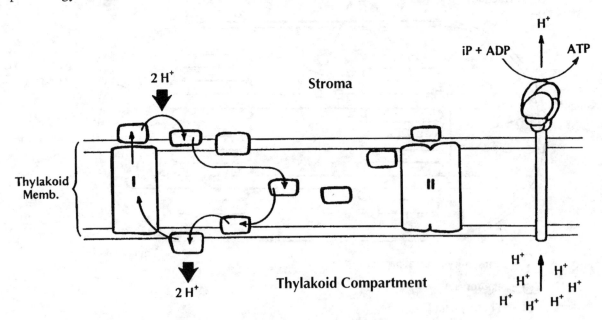

FIGURE 6-4. Cyclic Electron Flow

electric charge between the interior and exterior portions of the thylakoid disc represents an additional reservoir of energy. This gradient is what provides the energy that drives the synthesis of ATP.

A channel in a membrane embedded protein acts as a tunnel for the hydrogen ions to pass through the membrane. On the stroma side of the membrane, additional proteins are attached to the channel protein. These proteins catalyze the formation of a terminal phosphate bond on ADP, therefore synthesizing a high energy molecule of ATP. As the hydrogen ions rush back through the membrane to the stroma compartment to restore equilibrium, energy is released from the gradient. It is this energy that drives the formation of ATP.

What makes ADP and ATP so important to biological systems? It is their ability to capture and transfer, respectively, energy in chemical reactions. ADP is a molecule that can couple to energy-liberating (exergonic) reactions and capture some released energy in the formation of an additional phosphate bond, thus producing a high energy molecule of ATP. In turn, molecules of ATP can couple to energy-requiring (endergonic) reactions, and therefore donate some energy stored in their terminal phosphate bonds to drive the reactions. The energy is donated by breaking the terminal phosphate bond with the aid of a water molecule. It is the ability of ATP to supply energy to energy-requiring reactions that makes it a molecule of paramount importance to organisms. If there is an energy-requiring reaction taking place in an organism there is a high probability that ATP is supplying the energy to drive it.

The immediate importance of the production of ATP to the photosynthetic process will shortly be evident. However, we must postpone that discussion for a few minutes until we talk about the production of the other major product of the light reactions—**NADPH**.

Usually the photosystem (designated Photosystem I) associated with Cyclic Electron Flow operates in concert with another photosystem complex. This additional pigment complex is called **Photosystem II**. The two photosystems create a pathway called **Noncyclic Electron Flow**. The pathway is so named because the electrons do not cycle back to chlorophyll. The function of the pathway is to produce ATP and NADPH.

Noncyclic Electron Flow is initiated by having Photosystem II excited by sunlight, Figure 6-5. Electrons are raised from their "ground state" and passed to an electron acceptor molecule embedded in the thylakoid membrane adjacent to the system. Just as in Cyclic Electron Flow, the electrons pass through a series of acceptor molecules via oxidation/reduction reactions to the inner surface of the thylakoid membrane. As they travel from the outer to the inner membrane surface they carry along protons extracted from the stroma compartment of the chloroplast. When the electrons, with their protons in tow, reach the last acceptor in this chain, they liberate the protons to the inner watery compartment of the thylakoid disc. At the same time, Photosystem I is stimulated by light energy, and a pair of electrons is lost by the system. Immediately, the electrons from Photosystem II enter Photosystem I thereby eliminating the electron deficit of Photosystem I. The electrons from Photosystem I are grabbed by the same initial acceptor that functioned in Cyclic Electron Flow, but it is at this point that the two pathways begin to differ. The electrons, rather than being passed back through the membrane to the interior are used to reduce $NADP^+$. When $NADP^+$ accepts two electrons, it simultaneously extracts a proton from the stroma. This has the effect of lowering the hydrogen ion concentration, and producing a molecule of NADPH.

Just like Cyclic Electron Flow, Noncyclic Electron Flow activity establishes a hydrogen ion gradient that can generate ATP. The manner in which the ATP is made is identical to that of Cyclic Electron

Flow. The same phosphorylating apparatus is used by both Photosystems. Having the two photosystems act together produces both of the products of the light reactions—ATP and NADPH.

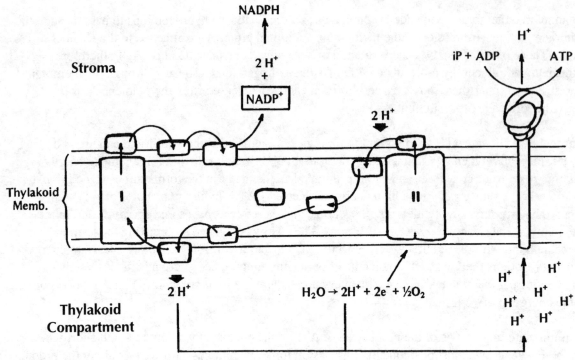

FIGURE 6-5. Noncyclic Electron Flow

Before we deal with the use of the energy in ATP and NADPH to make highly organized glucose molecules, we must clear up one detail of Noncyclic Electron Flow. That is, if Photosystem II contributes electrons to Photosystem I, then system II must be left with a deficit of electrons. How is this deficit corrected? Note from Figure 6-5 that when a molecule of water is split, it provides a source of electrons that can flow into photosystem II. Also note that it produces protons that contribute to the hydrogen ion gradient, and that oxygen is liberated as a gaseous by-product. Therefore, the source of oxygen during photosynthesis is from water not from carbon dioxide as was long believed.

The Dark Reactions

The two major products of the light reactions, ATP and NADPH, are ready to be used as fuel to drive the machinery of the dark reactions. The dark reactions are principally concerned with the creation of molecular order with the concomitant preservation and storage of energy. Specifically, the reactions reduce CO_2 by adding hydrogens from NADPH + H^+. ATP and NADPH supply the necessary energy for the reduction of carbon dioxide to glucose.

The creation of highly ordered glucose molecules is not the result of a hodgepodge of separate randomly occurring reactions, but the result of a logical series of reactions that build the glucose molecules in a stepwise fashion. It is not necessary for you to understand the intricate details of this complex metabolic pathway, but it should be possible for you to grasp a general understanding of how, through an orderly process and with the expenditure of energy, glucose is formed.
The dark reactions begin by taking advantage of the molecular order already established in the metabolic pathways inside the chloroplast. That is, a collection of CO_2, ATP, and NADPH does not

haphazardly react to form a six-carbon molecule instantly. Rather a series of metabolic reactions are linked that gradually adds molecular order to existing order. How this is accomplished is depicted in an abstract form in Figure 6-6.

Notice that the process begins with a 5-carbon molecule. Remember that the function of the dark reactions is ultimately to yield a 6-carbon molecule. To manufacture a 6-carbon molecule, it seems that by adding one carbon to the initial 5-carbon compound we'd be home free. But wait! If the pathway starts with 5-carbon compounds, wouldn't the supply eventually run out? The answer is obviously "yes," if nothing is done to replace the supply. What's the solution?

The answer is that the dark reactions ultimately manufacture 5-carbon replacement molecules besides producing a molecule of glucose. The specific manipulations of the molecules are complex and of no immediate concern to our discussion, but a summary of the major events of the process follows.

The first event in the dark reactions is the attachment of a molecule of CO_2 to a 5-carbon molecule. Before this can occur, a molecule of ATP must be invested in the 5-carbon molecule to make it a high energy compound. Once some energy from ATP has been transferred to the 5-carbon molecule, in the form of a phosphate bond, a molecule of CO_2 can react spontaneously with the energized molecule to form a 6-carbon compound. Almost immediately the 6-carbon compound breaks apart to form two 3-carbon compounds. At this point NADPH contributes hydrogens (really electrons in disguise) to the two 3-carbon compounds. This reaction requires the additional expenditure of ATP. The result of this reduction reaction is the formation of two high energy 3-carbon molecules. For lack of a better name, we will refer to these high energy 3-carbon compounds as G3P (an abbreviation of their chemical name that is not germane to our general discussion). If the preceding process is repeated five more times, it is possible to produce a total of 12 G3P molecules from six 5-carbon molecules and six molecules of CO_2.

It should be obvious that by this process six additional carbon atoms have been added to the original molecular order of the plant cell, i.e., to the six 5-carbon molecules. Although the six new carbon atoms are not directly associated with each other, they have nevertheless been added to the molecular order of the plant cell. Therefore, a newly synthesized glucose molecule does not consist of six newly incorporated carbon atoms.

One of the main accomplishments of the dark reactions is the transfer of energy from ATP and NADPH for the construction of new chemical bonds that establish molecular order. Remember, it is not only the building of molecular order that is of central importance in the photosynthetic process, but also the energy reservoir associated with the chemical bonds of glucose. It is this store of energy in the chemical bonds of glucose that will be tapped by all organisms. They will use this energy to construct new and different forms of biological molecules, and to drive all of their activities.

In the exercise that follows, you will learn to apply some techniques used to investigate certain aspects of the photosynthetic process. This will include learning how to extract the photosynthetic pigments from plant tissue, and how to separate and identify the various pigments that make up the photosystems. You will investigate how the pigments act together, and separately, to absorb certain wavelengths in the visible portion of the electromagnetic spectrum. A demonstration of chlorophyll fluorescence will illustrate how the membranes of the chloroplast play an important role in the photosynthetic process.

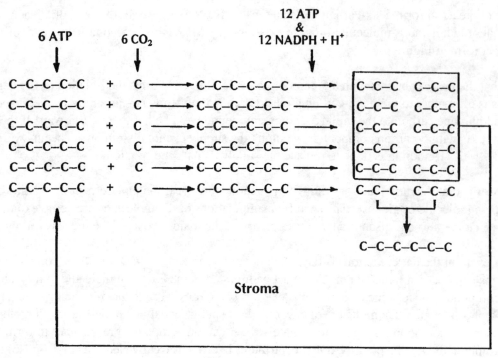

FIGURE 6-6. Summary of the Dark Reactions

OBJECTIVES

1. Be able to use and understand the concept of paper chromatography.
2. Be able to calculate R_f values for separated chloroplast pigments and to explain the concept of R_f values.
3. Determine an absorption spectrum for an extract of chloroplast pigments.
4. Determine individual absorption spectra for isolated pigments.
5. Understand how the individual absorption spectra of the isolated pigments relate to the spectrum produced by the chloroplast extract.
6. Understand the phenomena of absorbed, reflected, and transmitted light. (Why are plants green?)
7. Explain why extracted chlorophyll fluoresces when irradiated with a light source while chlorophyll inside chloroplasts does not fluoresce.

MATERIALS

chromatography paper
stapler and/or paper clips
developing jars
test tube racks
fermentation tube filled with pigment extract
black light for demo
diffraction grating (B&L Spectra Kit)
food coloring pigments

micropipettes or small capillary tubes
solvent—9:1 petroleum ether:acetone
Spec 20's
acetone
cuvettes
Pasteur pipettes
carousel projector and screen
pigment sample extracted with ethanol

acetone blanks
rulers
fresh plant tissue
scale
sand
graduated cylinder (25-ml)
funnel
cheesecloth (approximately 15 cm, two-ply)

scissors
plastic slide cuvettes
weighing paper
mortar and pestle
scoopula
beaker (100-ml)
filter paper
gloves (acetone impermeable)

I. EXTRACTION OF PHOTOSYNTHETIC PIGMENTS FROM PLANT TISSUES

Before you can begin to investigate the specific properties of photosynthetic pigments, you must first extract the pigments from the plant tissues without destroying their molecular structure. The photosynthetic pigments are organized into functional groups called **photosystems**, and are embedded in the membranes of the chloroplasts. Each functional group contains approximately 200 chlorophyll *a* molecules, 50 or so carotenoid molecules, a specialized molecule of chlorophyll *a*, and small amounts of chlorophyll *b*. Although isolated molecules have the capacity to absorb light, it is normally this entire complex of molecules that absorbs light during photosynthetic activity.

To extract the pigments, you must first physically break open the plant cells and liberate the chloroplasts. This is accomplished by grinding plant leaf tissue in a mortar and pestle. By grinding the tissues in acetone, you can simultaneously dissolve the pigments from the chloroplast membranes. The final step involves filtering the ground plant material through cheesecloth and filter paper. The final filtrate will be used for a number of the experiments that follow.

Procedure

General instructions—Three different kinds of plants will be available for you to extract pigment. Your instructor will assign groups of four to specific plants.

 CAUTION: Acetone is extremely flammable, use it only under the fume hood.

1. Pick several fresh leaves from the plant assigned to your group. On the scales provided, weigh approximately 5 grams of this tissue.
2. Cut the leaves into small bits with scissors, and transfer the pieces into the mortar.
3. Move the mortar to the fume hood. To the mortar, add a small scoop of sand and approximately 15 ml of acetone.
4. Grind the plant tissue in the mortar for 5 minutes, then allow the crushed sample to stand for an additional 5 minutes.
5. Prepare a filtration device by placing a 15 cm square of two-ply cheesecloth inside a filter paper-lined funnel, and position the funnel over a 100-ml beaker.
6. Wear gloves for the next few steps. Pour the contents of the mortar into the filtration device.
7. When the majority of liquid has passed through the filtration device, squeeze the tissue in the cheesecloth to collect any remaining liquid. Discard the cheesecloth and its contents in a container located in the fume hood.
8. The pigment extract that you have produced will be used for the paper chromatography separation of the individual pigments in the extract, and for colorimetric analysis to determine the absorption spectrum of the pigment extract.

Exercise 6 — Photosynthesis

II. SEPARATION OF PHOTOSYNTHETIC PIGMENTS

In this experiment, you will separate and isolate by partition chromatography the major pigments that make up the photosystems. The procedure is based on the fact that different organic molecules have different relative solubilities in various aqueous and nonaqueous solvents.

During the procedure, an extract of plant pigment is absorbed into a piece of chromatography paper. One edge of the paper is inserted into a solvent that migrates through the pigment deposit. Individual pigment molecules begin to dissolve and move in the (**non-aqueous**) solvent. However, the cellulose fibers of the paper contain bound water molecules that form a stationary (non-moving) aqueous phase; the solvent molecules form a mobile phase not soluble in the water phase.

Each type of pigment molecule has a different solubility in the solvent and water. Molecules that are highly soluble in the solvent but poorly soluble in water will be preferentially separated (partitioned) into the solvent phase and will move with the solvent. Molecules with a higher solubility in water will preferentially partition into the stationary water phase and will move at a slower rate. Each compound has a characteristic mobility (R_f) in a specific solvent system. R_f is the mobility of a compound relative to the migration of the solvent. That is, R_f is the ratio of the distance that a compound has migrated divided by the distance migrated by the solvent front. The observation that each compound has unique properties (R_f's in various solvent systems) has been applied to the purification and identification of many biological substances. If an unknown substance has the same R_f as a known (previously identified) compound in the same solvent system, the compounds are probably identical. Since individual compounds can be separated out of a mixture, chromatography has been used extensively to purify biological substances.

Procedure

1. Handle the chromatography paper only at the top. Oils from your hands can interfere with migrating pigment molecules.
2. Draw a light, horizontal pencil line (do not use ink) approximately 1.5 cm from the bottom edge of a strip of chromatography paper.
3. Use the micropipette that has the smallest opening to apply the pigment extract to the paper. Place the end of the pipette in the extract and then remove it. Some liquid should remain in the narrow portion of the pipette. Gently touch the tip of the pipette to the paper and simultaneously, in one smooth motion, draw a thin line across the paper. Use the pencil line as a guide. You will need to repeat this application procedure approximately 12-15 times before enough pigment is deposited. Before proceeding, ask your instructor if you have applied enough pigment.
4. Once the pigment is dry, bring the top corners of the paper together and staple or paperclip them to form a cylinder. Your chromatography paper should now stand in a vertical position without tipping over.
5. Place the chromatography paper in a developing jar (chamber) with the pigment line just above the liquid solvent. These chambers are in the fume hood and should not be removed from it. The solvent in the chambers is a mixture of 9 volumes of petroleum ether with 1 volume of acetone; both are extremely flammable, so do not smoke or have open flames in the lab. Do not allow the paper to touch the sides of the developing chamber or disturb the chamber once the chromatogram is inside. It may take 20 minutes or more for the chromatogram to develop. Do not let the solvent reach the top edge of the paper. While you are waiting for your chromatogram to develop, go to Part III.

6. When the solvent edge is approximately 1.0 cm from the top of the paper, remove the paper from the chamber. Quickly mark the solvent edge with a pencil mark. Allow the solvent to evaporate in the fume hood.
7. When the chromatogram is dry, identify the pigments separated. In ascending order, the pigments should be: chlorophyll b, chlorophyll a, xanthophylls, and carotenes. These four pigments are the major ones present. If minor pigment bands appear on your chromatogram, ignore them.
8. Calculate an R_f value for each of the pigment bands. Use the following formula:

$$R_f = \frac{\text{Distance a pigment migrated}}{\text{Distance the solvent migrated}}$$

9. Measure the distances along a common vertical axis to the tops of the pigment bands. Report the R_f values below.

Chlorophyll b $R_f = $ 6/9 = .067
(yellow-green)

Chlorophyll a $R_f = $ 1.7/9 = .19
(blue-green)

Xanthophylls $R_f = $ 3.8/9 = .42
(yellow—2 bands)

Carotenes $R_f = $ 9/9 = 1
(orange)

III. ABSORPTION OF LIGHT BY PLANT PIGMENTS

Not all wavelengths in the visible spectrum are used equally during photosynthesis. Plant pigments absorb light energy preferentially at specific wavelengths. Notice in Figure 6-7, that short wavelengths of radiation have a higher energy content than do longer wavelengths.

The extract that you will analyze is a combination of several photosynthetic pigments. Therefore, the amount of light absorbed by the extract is the sum of the light absorbed by the individual pigments present in the extract sample. You will use a spectrophotometer (Spec 20) to measure, at specific wavelengths across the visible spectrum, the relative amount of light absorbed by this combination of pigments. For a discussion of the theory and operating instructions of the Spec 20, consult Appendix II.

Procedure

1. At each Spec 20 station you will find a cuvette filled with acetone. The cuvette has a piece of para-film over the top to keep the acetone from evaporating (avoid inverting the tube). This cuvette will act as your blank for all spectrophotometer readings and should remain at the station at all times.
2. Fill a cuvette half full with acetone. To this tube add 10-15 drops of extract using a Pasteur pipette. You may need to adjust this amount depending upon the concentration you obtained during the extraction procedure. Mix the contents of the tube by inverting the tube (simply hold your finger over the end of the tube).

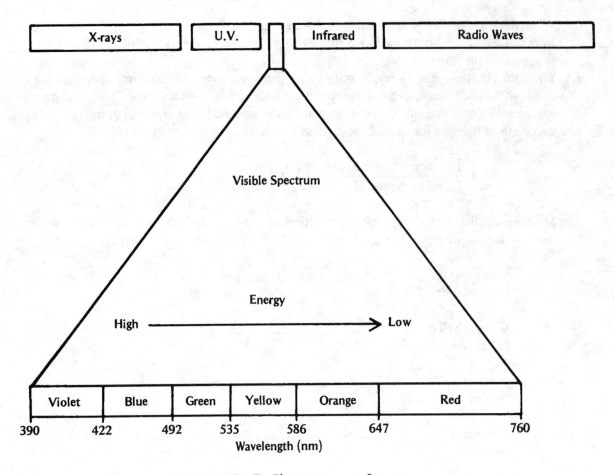

FIGURE 6-7. Electromagnetic Spectrum

3. Set the wavelength of the Spec 20 to 440 nm. Adjust the empty Spec 20 to register a transmittance of zero (left, front knob). Now, insert the blank and adjust the transmittance of 100% (right, front knob). Remove the blank and insert the sample tube. The absorbance of the sample should be approximately 0.70. If the absorbance value is too high, dilute the sample by adding acetone to it. If the absorbance value is too low, add more extract until the absorbance falls within the correct range. The purpose of this step is to provide a color range within which the Spec 20 can operate.
4. Remove the sample. Adjust the wavelength of the Spec 20 to 360 nm and set the transmittance to zero. Insert the blank and set the transmittance to 100%.
5. Remove the blank and insert the sample tube. Record the absorbance of the sample at this wavelength.
6. Remove the sample and set the wavelength at 380 nm. Reset the transmittance at zero (without the blank) and then insert the blank. Adjust the transmittance to 100%. Remove the blank, insert the sample tube, and read absorbance.
7. Repeat this procedure, recording absorbance values every 20 nm, up to a final wavelength of 680 nm. If you cannot reach 680 nm, stop when you can no longer set the machine to 100% transmittance.
8. Plot the absorbance values obtained as a function of wavelength. Use the graph on the next page (Figure 6-8).

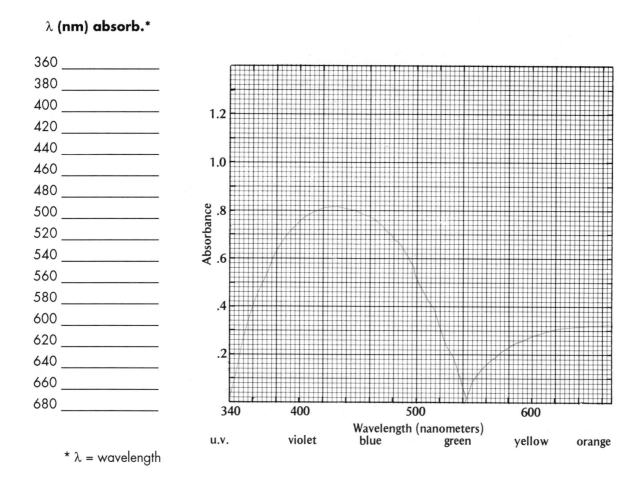

FIGURE 6-8. Graph for Plotting Absorption Spectrum of Chloroplast Extract

IV. A DEMONSTRATION OF CHLOROPHYLL FLUORESCENCE

Normally, chlorophyll and its associated pigments are organized into functional photosystems inside chloroplasts. They work together with the highly organized system of electron acceptor molecules that make up the remainder of the "Light Reaction" pathway. When photosystem pigments are extracted from leaf cells, the individual pigments become dissociated from the electron acceptor molecules.

In a nondisrupted chloroplast, electrons in high energy levels ("excited" electrons) are passed from the chlorophyll in the photosystems to electron acceptor molecules organized in a series of decreasing energy levels. During individual transformations to lower energy levels, relatively small amounts of energy are emitted. Some of this energy is used to make ATP. Gradually, the electrons resume their original energy levels by re-entering the photosystem (Figure 6-9).

Exercise 6 — Photosynthesis

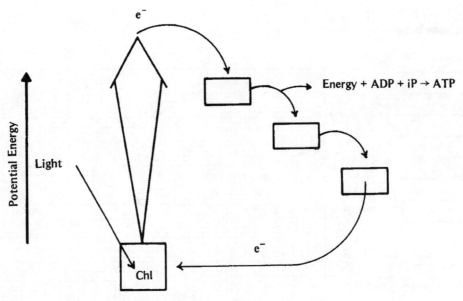

FIGURE 6-9. Acceptor Molecules Present in Intact Plant Cell
(iP = Inorganic Phosphate, Chl = Chlorophyll)

When chlorophyll pigments are isolated and irradiated with light energy, their electrons are raised to higher energy levels. Instead of returning to their original energy levels in gradual steps, the electrons fall rapidly and lose energy as light. This phenomenon is known as fluorescence (Figure 6-10).

Procedure

A small container filled with pigment extract will be irradiated with near-ultraviolet and black light radiation in a darkened room. What color is the fluorescence? Speculate about why fluorescence, in this case, is this color rather than another.

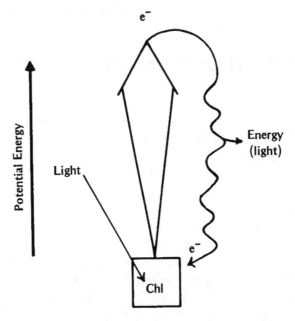

FIGURE 6-10. No Acceptor Molecules Present in Chlorophyll Extract
(Chl = Chlorophyll)

V. A DEMONSTRATION OF ABSORPTION BY VARIOUS PIGMENT MOLECULES OF SPECIFIC WAVELENGTHS IN THE VISIBLE SPECTRUM

White light contains all wavelengths in the visible range of the electromagnetic spectrum. When such light is passed through a solution containing a specific pigment, certain wavelengths can pass through while others are absorbed by the pigment molecules.

In this demonstration, white light will be separated and projected as the visible spectrum. Different colors of pigment will be inserted into the path of the white light before that light is separated into the spectrum (dispersed). The resultant spectrum will consist of wavelengths that were not absorbed by the pigment molecules.

Procedure

Different food coloring pigments will be separately inserted into a carousel projector. Complete and partially absorbed spectrums are projected. Why does blue food coloring appear blue? If you placed a sample of the pigment extract in the carousel projector, which color(s) would you expect to see projected? Why?

VI. ISOLATION OF SPECIFIC PLANT PIGMENTS AND A DETERMINATION OF THEIR INDIVIDUAL ABSORPTION SPECTRA—(if time permits)

In this exercise you will determine the relative amount of light absorbed by the individual photosynthetic pigments and determine at which wavelength(s) the absorption is maximal.

Procedure

1. Use the chromatograms prepared in Part II, "Separation of Photosynthetic Pigments." Begin by cutting the chromatograms with a scissors to separate the bands of pigments. (The double xanthophyll bands will be considered as a single band of pigment).
2. The class should distribute the isolated bands so that each group receives only one type of pigment band.
3. To place the isolated pigments back in solution (re-elute), simply put one or two of the isolated pigment strips into a cuvette and add a small amount of acetone. Shake the tube until all the pigment has dissolved in the acetone. Remove the paper strips. Add acetone to the tube until the volume is approximately the same as that in the blank tube.
4. To obtain a working concentration of pigment, follow the procedure of Step #3, Part III, "Absorption of Light by Plant Pigments." If the concentration of pigment is too low (below 0.7 absorbance) you will have to add another pigment strip to the cuvette to increase the pigment concentration.
5. Once you have a working concentration, determine the absorption spectrum of the isolated pigment. The procedure is the same as that used to determine the absorption spectrum of the extract sample.
6. Each group should place a graph of the absorption spectrum they obtained for their isolated pigment on the front board. The groups should be prepared to explain their results to the class. Record the individual spectra on the graphs provided on the following pages.

λ (nm)	absorb.*
380	_____
400	_____
420	_____
440	_____
460	_____
480	_____
500	_____
520	_____
540	_____
560	_____
580	_____
600	_____
620	_____
640	_____
660	_____
680	_____

* λ = wavelength

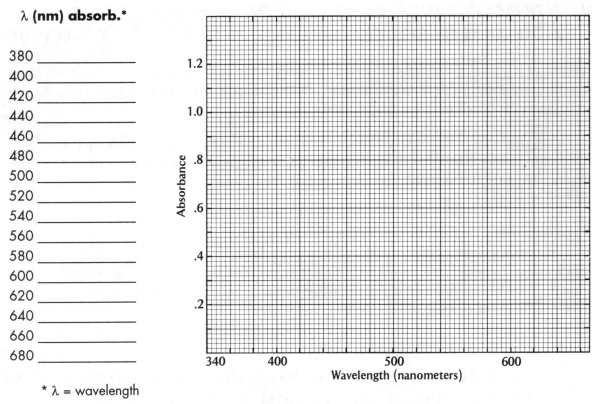

Graph for Plotting Absorption Spectra of Isolated Pigments

λ (nm)	absorb.*
380	_____
400	_____
420	_____
440	_____
460	_____
480	_____
500	_____
520	_____
540	_____
560	_____
580	_____
600	_____
620	_____
640	_____
660	_____
680	_____

* λ = wavelength

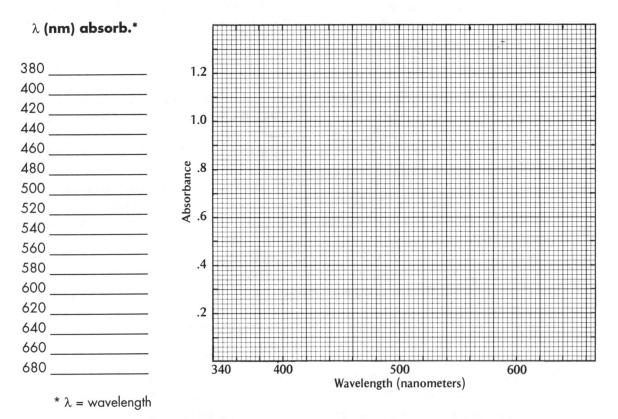

Graph for Plotting Absorption Spectra of Isolated Pigments

λ (nm)	absorb.*
380	_____
400	_____
420	_____
440	_____
460	_____
480	_____
500	_____
520	_____
540	_____
560	_____
580	_____
600	_____
620	_____
640	_____
660	_____
680	_____

* λ = wavelength

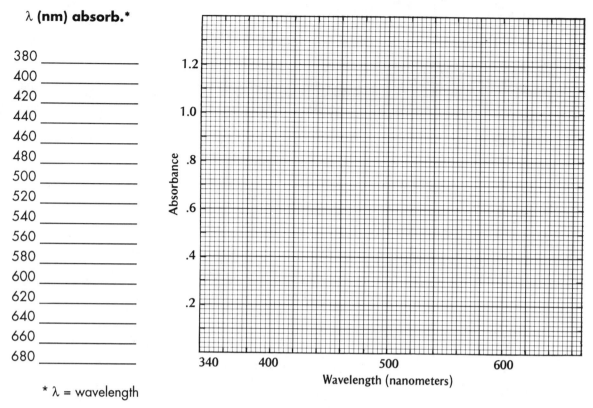

Graph for Plotting Absorption Spectra of Isolated Pigments

λ (nm)	absorb.*
380	_____
400	_____
420	_____
440	_____
460	_____
480	_____
500	_____
520	_____
540	_____
560	_____
580	_____
600	_____
620	_____
640	_____
660	_____
680	_____

* λ = wavelength

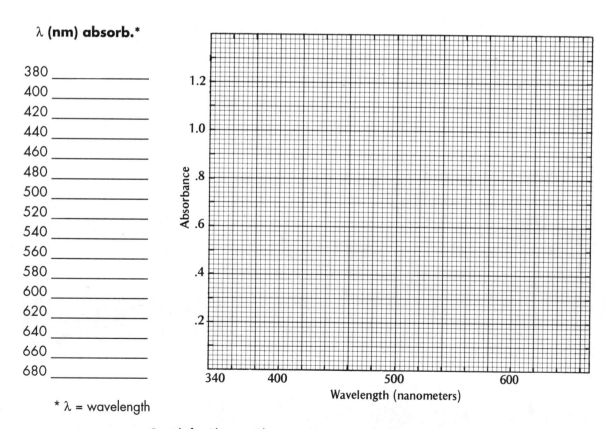

Graph for Plotting Absorption Spectra of Isolated Pigments

Exercise 6 — Photosynthesis

Mitosis and Meiosis

...cal events is that of reproduction. All organisms inherit the capacity ...luce their kind. The importance of reproduction should be intu- ...people appreciate the fact that it has been an ongoing process for ...ample, not one of your ancestors for as many generations as you ...luce and leave offspring. Those individuals that do not reproduce ...directly to the genetic makeup of their species. Therefore, the ...of individuals in a species have shaped and molded the evolution ...involves the transmission of hereditary information that conveys ...another cell or organism which generally resembles the parent cell

...kes place by a process called **binary fission**, meaning literally "to ...of most prokaryotes is contained in a single, long, circular strand ...chromosomes carry, in a linear arrangement, the basic units of ...in this group involves the replication of the genetic material and ...two copies in new daughter cells. Because the single chromosome ...the length of the parent cell, an orderly process for duplicating ...DNA molecules is required. For a detailed description of this ...extbook.

The ... reproduction in **eukaryotes**. Cell division is actually two disc... ...asmic. **Mitosis** and **meiosis** are the nuclear processes that invo... ...terial (**chromosomes**) into new nuclear areas. Mitosis is a conse... ...clear areas that contain the identical complement of genetic mater... ...cell. Meiosis in contrast, produces nuclear areas that contain only ... ...rent cell. Once the nuclear processes of mitosis and meiosis are compl... ...ivides the cytoplasm around these new nuclear areas.

In asex... ...duction of an individual is synonymous with cell division. In these o... ...ensures that both of the nuclear areas contain an exact copy of the gen... ...e parent cell. The process of cytokinesis partitions the cyto- plasm a... ...rming two new genetically identical cells. The process of meiosis ... ...nicellular eukaryotes.

In sexually reproducing eukaryotes, both types of cell division are present: mitotic and meiotic. Again, as in asexual unicellular organisms, the mitotic process of cell division, coupled with cytokinesis, involves the production of identical copies of cells that have constant numbers and kinds of chromosomes from generation to generation. In multicellular eukaryotic animals, the cells produced from mitotic cell division generally do not lead individual existences, but become part of a multicellular structure. Therefore, the mitotic process in these organisms is primarily responsible for building a multicellular organism from a single cell. With but a few exceptions, the millions or perhaps billions of cells that make up a multicellular organism have identical genetic constitutions, all due to the mitotic process. You will examine in some detail this amazing process that faithfully distributes the genetic material with seldom an error.

Generally, the **somatic** (body) cells of sexually producing multicellular organisms contain a constant number of chromosomes. For example, in humans each body cell contains 46 chromosomes. If the size and shape (**morphology**) of the 46 chromosomes are compared, it is apparent that they can be arranged as 23 pairs. That is, for every chromosome that has a particular size and shape, there would be another in the total complement that has a similar morphology. In addition, if the gene loci (positions) along the lengths of each chromosome of a pair are considered, it would be found that the loci are identical. In other words, if two genes at the same loci are compared, one from each chromosome, they would be identical in the sense that they influence the same trait. The two genes might influence the trait identically or they might affect the trait differently. Chromosome pairs that are morphologically similar and carry identical sequences of genes are called **homologous chromosomes**. Cells that have the complete complement of homologous chromosomes are called **diploid (2N)**.

How do cells get the two copies of each type of chromosome? Where did the individual copies come from? The answers to these questions are found in an explanation of sexual **reproduction** and the process of **meiosis**.

Diploid cells are half **maternal** and half **paternal** with respect to their chromosome complement. The sources of the maternal and paternal chromosomes are the **gametes**. Gametes are the reproductive cells produced by meiosis in animals and by mitosis in sexually reproducing multicellular plants. These gametes contain only half the total number of chromosomes, that is, they contain the **haploid (N)** number. In plants, haploid spores are obtained by meiotic cell division. Gametes produced by the female are called **eggs**; gametes produced by the male are called **sperm**. When male and female gametes unite, in a process called **fertilization**, they restore the chromosome number to the diploid condition. The single cell formed during fertilization is called the **zygote** (fertilized egg). It is the zygote and the subsequent cells produced from the zygote that undergo mitotic cell division to give rise to a multicellular organism.

In sexually reproducing multicellular animals, gametes are produced from special diploid cells called **germ cells** which undergo meiotic cell division. These germ cells are usually restricted to specific locations, the **gonads** (ovaries in females, testes in males). All of the cells other than the germ cells are called **somatic** cells. In multicellular sexually reproducing plants, the haploid spores are produced by meiosis from special diploid cells called **sporocytes**.

In the exercises that follow you will look more closely at the meiotic process that reduces the chromosome number in half to produce haploid cells.

Figure 7-1 represents a life cycle of a sexually reproducing multicellular animal and shows where meiosis, mitosis, and fertilization occur. Notice that the process of meiosis marks the end of the diploid phase of the life cycle and the beginning of the haploid phase. Similarly, the process of fertilization marks the end of the haploid phase and the beginning of the diploid phase. In this life cycle, mitosis is restricted to the diploid phase.

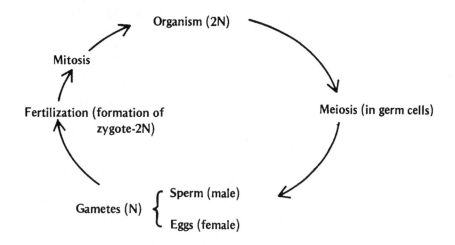

FIGURE 7-1. Life Cycle of a Sexually Reproducing Multicellular Animal

Figure 7-2 is a life cycle of a sexually reproducing multicellular plant. Plants have a wide range of life cycles with mitosis present in both the diploid and haploid phases. The products of meiosis, the haploid cells called spores, undergo mitotic cell division to produce multicellular structures. At a certain stage in the haploid phase, specific cells undergo mitotic cell divisions to produce eggs or sperm. Notice that the meiotic process is not necessary to form eggs or sperm because the cells that give rise to the gametes are already haploid. The diploid phase that gives rise to spores is called the **sporophyte**, while the haploid phase that gives rise to gametes is called the **gametophyte**.

Figures 7-3 and 7-4 illustrate that the phylogenetic group, the mosses, has a life cycle in which the gametophyte (haploid) is the dominant phase, with the sporophyte dependent upon the gametophyte for support and nutrition. In contrast, the flowering plants have completely reversed this role, so that the sporophyte (diploid) is the dominant phase, while the gametophyte is extremely small and restricted to living inside the tissues of the flower. Ferns are intermediate between the mosses and flowering plants, having the sporophyte dependent upon the gametophyte for only a short time during the life cycle.

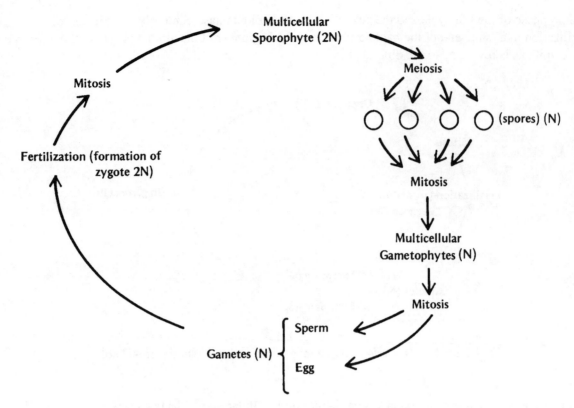

FIGURE 7-2. Life Cycle of a Sexually Reproducing Multicellular Plant

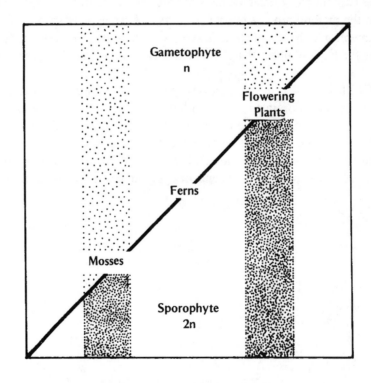

FIGURE 7-3. Dominance Relationship Between the Sporophyte and Gametophyte

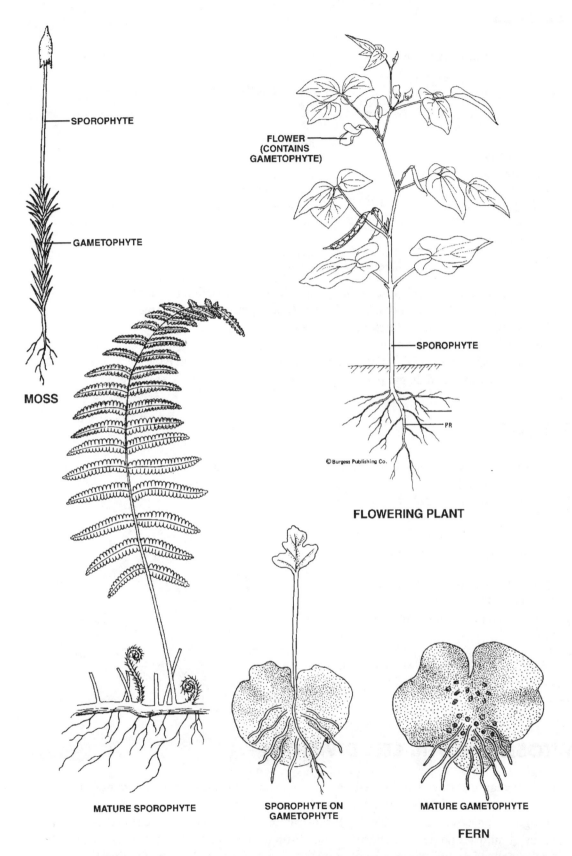

FIGURE 7-4. Moss, Fern and Flowering Plant Sporophytes and Gametophytes

OBJECTIVES

1. Define the following terms:

cell division	somatic cells
mitosis	spores
meiosis	sporophyte
cytokinesis	gametophyte
chromosome	chromatid
homologous chromosomes	fertilization
centromere	germ cells
diploid	spindle apparatus
haploid	meristematic region
gametes	reduction division
synapsis	tetrad

2. Draw the general life cycle of a sexually reproducing multicellular animal.
3. Draw the general life cycle of a sexually reproducing multicellular plant.
4. Explain the function(s) of mitotic cell division.
5. Explain the function(s) of meiotic cell division.
6. Be able to explain and diagram the major phases of mitotic cell division.
7. Be able to explain and diagram the major phases of meiotic cell division.

MATERIALS

chromosome models (red and yellow pop-it beads, Chromosome Simulation BioKit—Carolina Biological Supply)
2" x 2" color slides or digital images of mitotic figures
prepared slide of *Allium* root tip
Lilium sp. flower slides, X.S. (anther series)
2" x 2" color slides or digital images of meiotic figures
onion (*Allium*) sets
dropper bottle with aceto-orcein stain (45%)
microscope slides and coverslips
spot depression plate with cover glass
incubator
razor blades
forceps
disposable pipettes
white string

I. MITOSIS (MITOTIC CELL DIVISION)

In the exercises that follow, you will investigate the series of events that characterize mitotic cell division. Chromosome models will be used in the first exercise to trace the movement of chromosomes through the phases of mitosis. Next, your instructor will point out, from colored slides, the major phases of mitosis and the characteristics that distinguish each. Then you will use prepared slides of cells stopped in different phases of mitosis and cytokinesis to become familiar with the steps of mitotic cell division. Mitotic cell division may be broken down into the following descriptive stages: **Interphase, Prophase, Metaphase, Anaphase,** and **Telophase**.

A. Chromosome Models—Mitosis

Work in groups of four. Each group should have a jar of red and yellow "pop-it" beads along with at least eight plastic-covered cylindrical magnets. Individual beads assembled in a string will represent the linear arrangement of genes on a chromosome. The centromere, an area were the chromatids of a chromosome remain attached until division, will be represented by a magnet.

Procedure

1. Assemble the red pop-it beads in strings. Do not alternate the color of the pop-it beads in the strings! Attach a string of beads to each end of the magnet. Make another model identical to the first, that is, one with the same color and number of beads. Assemble two more chromosome models out of the yellow beads. Make the models identical to the two previously constructed red models; the only difference should be the color.

2. Place two pieces of white string on the top of the lab table. The string represents two uncoiled chromosomes as they would appear during the initial stage of cell division, **interphase**. Information is being translated, proteins synthesized, and the genetic material duplicated. Cells in interphase have intact nuclear membranes.

3. When the cell has completed chromosome duplication, it is ready to enter the first phase of mitosis (nuclear division), **prophase**. During prophase, the nuclear membrane disappears and the diffused chromosomes begin to coil up. Not until the genetic material condenses do the chromosomes become morphologically recognizable as **chromosomes**. Because the genetic material is duplicated during interphase, each visible chromosome contains two identical copies of the original chromosome. These identical strands of genetic material are called **chromatids**. Chromatids can be thought of as "future chromosomes." They become chromosomes only after they have separated from each other.

4. Connect the identical models together at their centromeres. You should have two yellow models hooked together and two red models attached to each other. The models attached to each other represent the chromatids of a chromosome. Replace the string with the chromosome models. As prophase progresses, the chromosomes become shorter and thicker. While the chromosomes are condensing, the **spindle apparatus** is forming. The spindle apparatus consists of spindle fibers which are hollow microtubules. The arrangement of the spindle fibers determines the direction of movement of the chromosomes. At the end of prophase the chromosomes are in their maximum condensed state and their arrangement is essentially random.

5. The next phase, **metaphase**, begins as the chromosomes start to move toward the equatorial plate of the cell. Remember that each chromosome is made up of two chromatids. The end of metaphase is marked by the arrival of all chromosomes at the equatorial plate. Move your models so that they form a line. You should have two chromosomes (one red and one yellow), each composed of two chromatids, in the line. Perpendicular to this line and on both sides would be two polar regions from which the spindle fibers radiate.

6. The beginning of **anaphase** is initiated by the separation, in each chromosome, of the chromatids at the centromere region. Once the centromeres have separated, the chromatids (now individual chromosomes) are free to move toward opposite poles.

   At this time, separate the magnets and move the half-models to opposite sides of the imaginary equatorial plate. Notice that on each side you have one red and one yellow chromosome model, not models of one color. This complement of chromosomes is exactly what the parent cell originally possessed. Anaphase is complete when the chromosomes have reached the poles.

7. During the final phase, **telophase**, the chromosomes begin to uncoil, the spindle apparatus disappears, and the nuclear membrane reforms. Division of the cytoplasm (cytokinesis) occurs during this phase. In animal cells, cytokinesis involves the pinching in of the cytoplasm from all sides. This creates a deep furrow that eventually isolates the two nuclear areas and their respective cytoplasm.

   Because plants have a rigid cell wall, "furrowing" is not possible. Instead, a cell plate is formed at the old equatorial line. This cell plate eventually develops into a barrier that separates the cytoplasm into two parts. The end of cytokinesis marks the completion of cell division. The end products of mitotic cell division are two genetically identical daughter cells. Figure 7-5 summarizes the events of mitotic cell division.

B. Phases of Mitosis Illustrated by Computer Graphics—Instructor Demonstration

C. Cell Division In The Root Tip Of Onion (*Allium*)—Student and Commercially Prepared Slides

In higher plants, mitotic cell division is restricted to specific areas of the plant body called **meristematic regions**. Two major areas of meristematic tissue are found in shoot and root tips (the **apical meristems**). You will prepare your own slides of onion cells from the meristematic region of root tips, and use these slides to find representative stages of mitosis and cytokinesis. The method of slide preparation is called a **squash technique**, and the procedure is described below. You will also use commercially prepared longitudinal sections of onion root tips to observe mitotic figures. The meristematic region in these sections is located just posterior to the root cap (Figure 7-6). Sketch each stage of mitotic cell division in the spaces provided below, and list the distinguishing characteristics of each stage.

Procedure For Squash Technique

1. Each group of four students should obtain an onion that has a healthy group of young roots. There should be enough root tips per onion to provide each group with 8-12 slides.

2. Using a forceps, detach the distal 1 cm of each root tip. Place the root tips in a single well of your depression plate. Next, obtain a bottle of aceto-orcein stain from the incubator. With a dropper add enough stain into the well to completely cover the roots, filling the well about 1/2 full. Place the stain bottle back in the incubator.

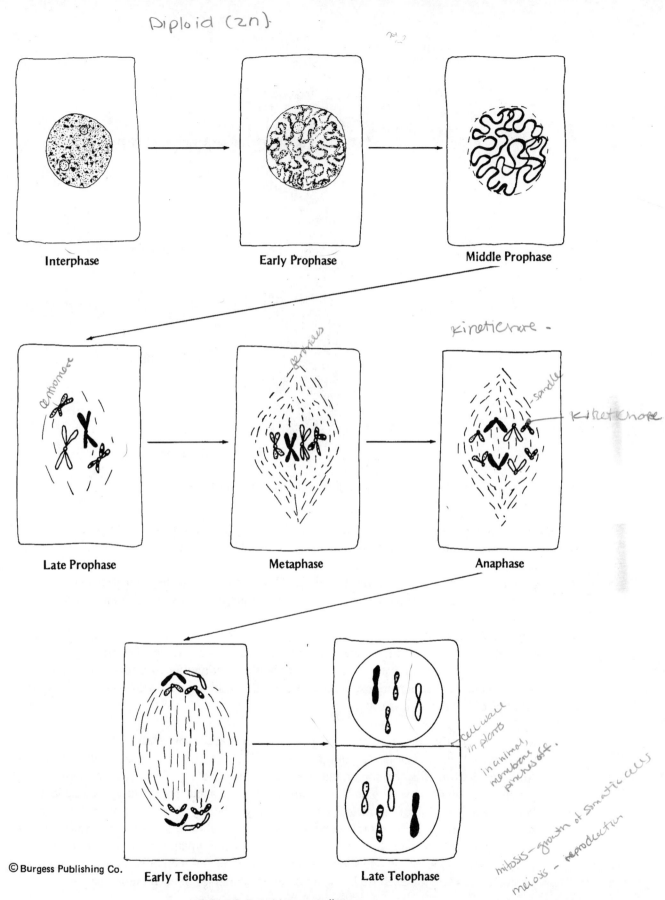

FIGURE 7-5. Mitotic Cell Division

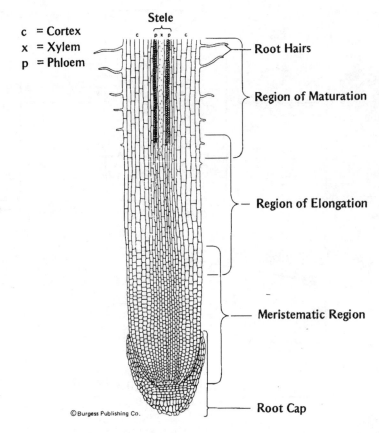

FIGURE 7-6. Longitudinal Section of an Onion Root Tip

3. Label your depression plate so it may be identified by your group. Cover the plate with the glass cover to prevent evaporation and place the plate in the incubator. The incubator should be set to approximately 60°C. Allow the roots and stain to incubate for 45 minutes. The acetic acid in the stain mixture will fix the cells and break down the adhesive material (middle lamella) between the cells. Orcein is an excellent stain for nuclei and chromosomes.

4. Remove the depression plate from the incubator, and suction off the stain using a disposable pipette. In a steady stream of water, discard the used stain down the sink. To rinse the root tips, refill the well with water from a dropper bottle. Suction this water out of well and discard in the same manner described above. Again, refill the well with water and leave the root tips covered with this water.

5. Remove a root tip from the depression plate with a tweezers, and place it on a clean microscope slide. With a razor blade, cut off the terminal 2 mm of the root tip, and discard the rest. Add a small drop of water, and carefully add a plastic coverslip. With the eraser end of a pencil, press on the coverslip until the cells of the root tip are spread out in a single layer. With a paper towel, blot the surface of the slide dry. Check your progress with the microscope. If you do not have a single cell layer, remove the slide and again apply pressure to the coverslip. Return the slide to the microscope. Using a combination of your slides (you may want to prepare more than one slide per person), and the commercially prepared longitudinal sections, complete the exercise by sketching representatives of each mitotic phase.

Stages of Mitotic Cell Division	Distinguishing Characteristics
Interphase	1. chromosomes are uncoiled 2. intact nuclear membrane 3. info being translated, proteins synthesized & genetic material duplicated
Prophase	1. nuclear membrane disappears 2. genetic material condenses 3. chromatids form (late) spindle apparatus forms
Metaphase	1. Chromatids line up in the metaphase plate
Anaphase	1. Chromatids seperate at the centromere 2. move towards the poles
Telophase	1. Chromosomes uncoil 2. spindle disappears 3. nuclear membrane reforms 4. cytokinesis (division of cytoplasm) animals - pinch off, plants - cell wall

II. MEIOSIS (MEIOTIC CELL DIVISION)

Meiotic cell division results in the production of cells that have the haploid number of chromosomes. Meiosis is a nuclear division that reduces the diploid number of chromosomes to the haploid number. The discussion of the meiotic process will be restricted to the higher plants and animals. The same format used to explain mitosis, with models, colored slides, and prepared slides, will be used to explain meiosis.

In the previous mitotic chromosome model exercise, the hypothetical cell you worked with had one pair of homologous chromosomes, therefore the cell was diploid. To demonstrate the movement of chromosomes through the meiotic process with the subsequent production of haploid cells, you will use a cell with the same diploid number of chromosomes, two.

Meiotic cell division may be broken down into the following descriptive stages: **Interphase, Prophase I, Metaphase I, Anaphase I, Telophase I, Interkinesis, Prophase II, Metaphase II, Anaphase II,** and **Telophase II**.

A. Chromosome Models—Meiosis

Again work in groups of four. Use the same chromosome models that you constructed for Part I.

Procedure

1. **Interphase** in both mitotically and meiotically dividing cells is essentially identical, that is, the genetic material is translated and duplicated, and protein is synthesized.

2. During early **prophase I**, the uncoiled genetic material begins to coil up into recognizable chromosomes. Remember that when the chromosomes appear during prophase I, replication has already occurred. The spindle apparatus begins to form and the nuclear membrane disappears during this phase. Up to this point the phases of mitosis and meiosis are essentially identical.

    The first significant difference between the two processes occurs later in prophase I with the pairing (**synapsis**) of the homologous chromosomes. The homologous chromosomes lie near each other along their length. The resulting structure is called a **tetrad**; it is composed of four chromatids (two chromatids from each chromosome of a homologous pair). To form a model tetrad, attach the homologous chromosome models to each other by their magnetic centromeres. When synapsis has occurred, prophase is complete.

3. **Metaphase I** begins with the movement of the synapsed homologous chromosomes to the equatorial line of the cell. Metaphase I is complete once the tetrads are at the equator. Notice that the homologous chromosomes do not line up independently at the cell equator. Move your tetrad model to the imaginary cell equator.

4. The major difference between mitosis and meiosis occurs during **anaphase I**. It is during this phase that the actual reduction in chromosome number takes place. Anaphase I is therefore called the phase of **reduction division**.

Anaphase I starts with the separation of the homologous pairs at the equatorial line. The centromeres of the individual chromosomes remain intact; it is the synaptic junction between the two homologous chromosomes that breaks. Once the homologous chromosomes are separated, they are free to move toward opposite poles. When the chromosomes reach the separate poles, anaphase I is complete.

Separate your tetrad model in a way that allows you to move one red chromosome to the left and the yellow model to the right. Each chromosome should be composed of two chromatids of the same color.

5. **Telophase I** is essentially identical to telophase of mitosis except that the daughter cells produced are haploid, not diploid as they are in mitosis. The chromosomes in these haploid daughter cells are double stranded (two chromatids), whereas the chromosomes in daughter cells produced during mitosis are single stranded (only one chromosome). During telophase I, the spindle apparatus disappears, the chromosomes uncoil, the nuclear membranes reform, and the cytoplasm divides.

6. During interkinesis, the genetic material does not replicate.

7. Each daughter cell undergoes a series of events that are fundamentally the same as mitosis. In **prophase II**, the genetic material recondenses into visible chromosomes. Each chromosome is made up of two chromatids. Recall that the daughter cells produced during the reduction division are haploid, therefore the chromosomes in haploid cells no longer have their homologous counterparts. Consider the chromosome models to be isolated in separate cells.

8. In **metaphase II**, the chromosomes form a line at the equatorial plate.

9. The centromeres of the chromosomes divide at the beginning of anaphase II. This allows the chromatids (now chromosomes) to separate from each other and move toward opposite poles. Anaphase II is complete when the chromosomes reach the poles.

Separate the chromatids in each of the imaginary haploid daughter cells and move them toward opposite poles.

10. The events of **telophase II** are basically the same as those of mitotic telophase. During telophase II, the chromosomes uncoil, the spindle apparatus disappears, the nuclear membrane forms, and the cytoplasm divides. Notice that the new daughter cells retain the haploid number of chromosomes and do not assume a diploid condition though they were produced from an essentially mitotic division (anaphase II).

Figure 7-7 summarizes the events in meiotic cell division.

11. Build two more homologous chromosome models, each composed of two chromatids, but place the centromere in a different position and vary the length from that of the existing models. Once you've constructed the models, manipulate them through the different phases of meiosis. Note: You will be starting meiosis with a hypothetical cell that has four chromosomes instead of the two chromosomes of the previous example.

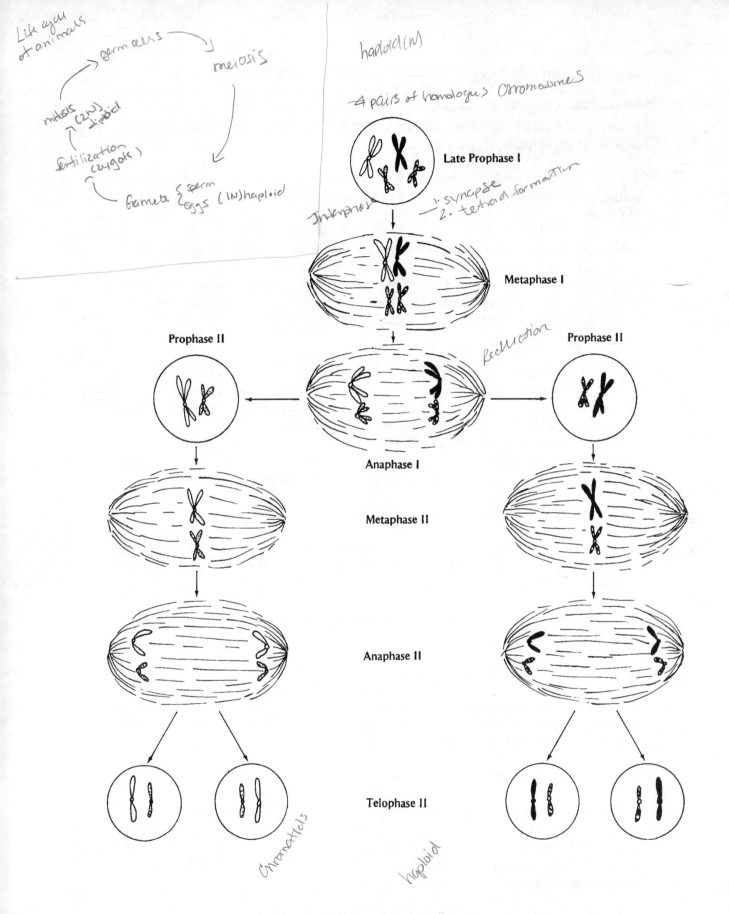

FIGURE 7-7. Meiotic Cell Division

B. Phases of Meiosis Illustrated by Computer Graphics—Instructor Demonstration

C. Meiotic Cell Division in the Lily Flower (*Lilium*)—Prepared Slides

Just as the mitotic process is restricted to the meristematic areas, meiotic cell division in flowering plants is restricted to specific regions of the diploid plant body. The meiotic process occurs only in organs called **flowers**. In certain species, the tissues that undergo meiosis to produce male and female haploid spores are found in separate flowers, while some flowers of other species possess both tissues. The tissue that gives rise to male haploid spores is found in the **anther** of the flower, while the **ovary** contains the tissue that produces the female haploid spores.

Look at prepared slides of the lily flower (Figure 7-8). Cross sections of the flower include both anther and ovary structures. Ignore the ovary and concentrate on the meiotic phases that occur in the anther. Sketch the phases of meiosis listed on the next page from prepared slides and list the distinguishing characteristics of each phase.

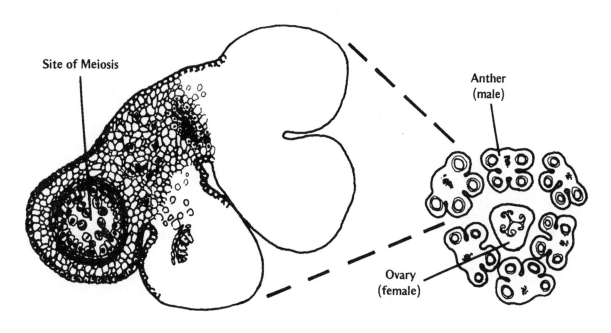

FIGURE 7-8. Cross Section of a Lily Flower

Stages of Meiotic Cell Division **Distinguishing Characteristics**

Interphase (respiration)

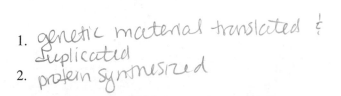

1. genetic material translated & duplicated
2. protein synthesized

Stages of Meiotic Cell Division | **Distinguishing Characteristics**

Early Prophase I
genetic material begins to coil up

1. spindle forms
2. nuclear membrane disappears

Late Prophase I

1. pairing of homologus chromosomes (tetrad)
2. 4 chromatids

main difference from mitosis → synapses - exchange of genetic info — centromere loci

Metaphase I (1st Division)

1. homologus chromosomes to equator by spindle fibers
2. don't line up independently

Anaphase I   1. seperation of homologus pairs   2. reach seperate poles

Telophase I (haploid)
double stranded

1. spindle disappears
2. chromosomes uncoil
3. nuclear membranes reform
4. cytoplasms divide (cytokinesis)

Metaphase II (2nd Division)

1. chromosomes form line at equatorial plate

Late Anaphase II

1. centromeres split
2. move towards the poles

Telophase II *(haploid gametes)*

1. chromosomes uncoil
2. spindle disappears
3. nuclear membrane forms

(For lily flower)

8 Genetics

AaBbCcDdEeFfGg → how many gametes? 2^n
n = heterozygotic alleles = 3 $2^3 = 8$

INTRODUCTION

Why do members of a sexually reproducing species resemble each other from generation to generation? More specifically, why do offspring resemble their parents in the characteristics that make them appear morphologically similar, but at the same time differ with respect to certain other characteristics? The answers to these questions are found in the study of genetics, the branch of biology that deals with inheritance.

The units of inheritance, the **genes**, are segments of a long, coiled strand of DNA called a **chromosome**. A typical chromosome in humans may contain thousands of different genes. The position of a gene along the length of a chromosome is called its **locus**. In each diploid cell of an organism, there can be a maximum of only two loci for the same gene (a gene pair), one on each chromosome of a homologous pair of chromosomes. **Homologous chromosomes** have the same linear arrangement of genes along their lengths. The two loci for a gene may be identical, that is, they may influence a characteristic in the same manner, or they may influence the characteristic differently from one another. In this latter case, these different forms of the same gene are referred to as **alleles**. For a given type of gene in a diploid cell, there can be a maximum of only two alleles, but a population of organisms can harbor many forms of the same gene. These different forms of the same gene in a population are referred to as **multiple alleles**.

Not all alleles influence a characteristic in an equal manner. When a cell contains alleles for a given trait, one allele may be **dominant** over the other. A dominant allele can determine or influence a characteristic when one or both chromosomes of a homologous pair possess the allele. Therefore, in order for a dominant allele to be expressed, it is only necessary that one chromosome of a homologous pair possess that allele. In contrast, a **recessive** allele must be present on both chromosomes of a homologous pair if it is to determine or influence a characteristic. Capital letters are usually used to designate a dominant allele (B), and lowercase letters for a recessive allele (b). The choice of letters to indicate a particular kind of gene is arbitrary. When considering dominant and recessive alleles, a diploid cell has three possible allelic combinations. They are: both chromosomes may possess dominant alleles (BB), both chromosomes may have recessive alleles (bb), or the chromosomes may carry different alleles (Bb). The condition in which the alleles are the same, regardless of dominance or recessiveness, is called **homozygosity** (BB or bb). When there are two different alleles present, the condition is called **heterozygosity** (Bb).

Consider the example of two alleles, one that produces black hair (B) and one that produces brown hair (b). The black allele is completely dominant over the brown allele. For the particular characteristic in question (color), the letter designations BB, Bb, and bb indicate the genetic makeup (**genotype**) of an organism. The genotype indicates what kinds of alleles are present and provides the necessary information to tell whether an individual is homozygous or heterozygous for a trait. How a characteristic appears to our senses is called the **phenotype**. Therefore, even though BB and Bb are two different genotypes, they have the same phenotype—black. Figure 8-1 illustrates the possible genotypic combinations and their respective phenotypes.

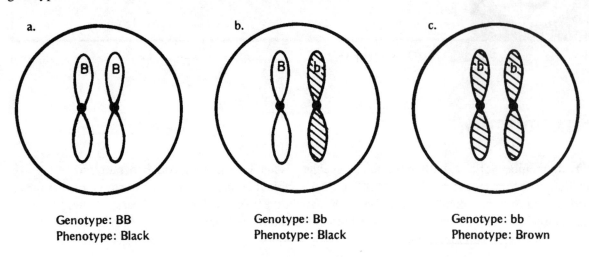

FIGURE 8-1. Genotypic Combinations and Respective Phenotypes for a Hypothetical Dominant/Recessive Trait of Black/Brown color

Remember, in sexually reproducing organisms, the process of **meiosis** is one of the initial steps in the transmission of genetic material (chromosomes) to offspring. It is during this process that the homologous chromosomes s**eparate** and **segregate** (isolate) into haploid cells. The alignment of the synapsed homologous chromosomes at the equatorial plate during metaphase I, and their subsequent separation and movement in opposite directions, is a **random** event. Therefore, the direction of movement of the members of one homologous pair is totally independent of the direction of movement of the members of all other homologous pairs. In the haploid cells produced during meiosis, only one chromosome from each homologous pair is present. If a diploid cell, heterozygous for a particular gene, undergoes meiotic cell division to produce haploid gametes, then two types of gametes are possible. For example, if the cell illustrated in Figure 8-1b undergoes meiotic cell division, four haploid cells will be produced, two with chromosomes carrying the dominant allele B and two with the recessive allele b.

If you don't understand how the process of meiosis generates different types of gametes from diploid cells with different genotypes, you should go back and review meiosis. Being able to generate gamete types from different genotypes is usually one of the first steps in solving Mendelian genetics problems.

When a diploid cell has many homologous pairs of chromosomes, the number of different chromosome combinations that can possibly be produced during meiotic cell division is extremely large because of **random segregation**. The number of possible chromosomal combinations in the haploid cells produced during meiosis from a diploid cell is equal to 2^n, where n is the number of different types of chromosomes, usually the haploid number. In humans for example, the number of different chromosomal combinations is 2^{23} or 8,388,608. If you add to this variation the amount produced from crossing-over, the number of potentially different gametes becomes astronomical, 2^{2017}. To put this in perspective, the number of electrons in the entire universe is estimated to be 2^{80}.

Therefore, the random segregation and independent assortment of chromosomes into gametes is an important source of genetic variation in populations.

The next stage in the transmission of genetic material to offspring is the reuniting of the homologous chromosomes during fertilization. Only when the chromosomes from the homologous pairs have been separated and isolated into gametes can this occur. During fertilization, each gamete contributes one chromosome from each homologous pair. The fusion of the gametes produces a diploid cell (**zygote**) that has its genetic makeup determined by the chromosome types present in the gametes. Therefore, the number of different kinds of gametes a female and a male can produce will ultimately determine the kinds of offspring, and the relative numbers thereof, that can be produced. Assuming that each parent has the potential to produce 2^{23} different gametes, how many different types of zygotes could two human parents potentially produce? (Ignore the additional source of variation produced during crossing-over).

The reuniting of chromosomes during fertilization is, like segregation and independent assortment during meiosis, a random (chance) event. The chance of a sperm uniting with one kind of egg cell, when two types of egg cells are present in equal numbers, is 50%. This also holds true for a particular egg type if two sperm types are present in equal numbers.

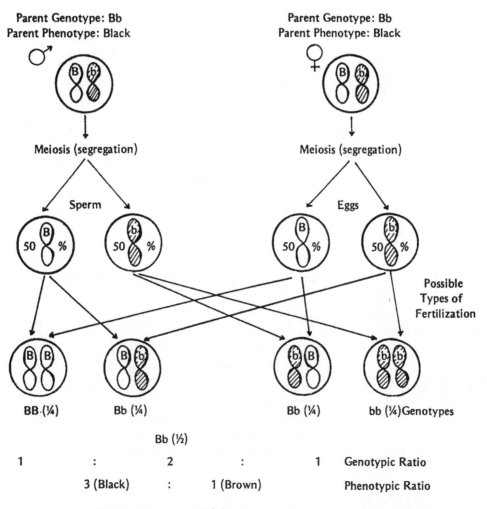

FIGURE 8-2. A Summary of the Events Leading to the Transmission of Genes to Offspring If a Single Pair of Alleles is Considered

Figure 8-2 summarizes the events leading to the transmission of genes to offspring if two alleles are considered. The illustration uses the hypothetical dominant/recessive trait of black and brown color. Notice both parents produce two types of gametes with equal frequency. Also notice a particular gamete has an equal chance of uniting with either of the two different types of gametes produced by the opposite sex. Figure 8-3 illustrates a much more convenient method for determining patterns of inheritance using a Punnett square using a genetic shorthand without illustrating the chromosomes or sex. The essential information in Figure 8-2 is also present in Figure 8-3.

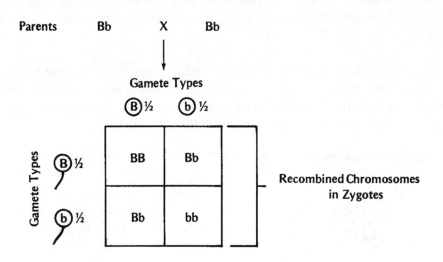

FIGURE 8-3. Summary of Figure 8-2

PRELAB PREPARATION

Before you come to lab you should be familiar with the following terms and processes:

chromosome	dominant allele
homologous chromosomes	recessive allele
allele	autosomes
homozygous	sex-linked genes
heterozygous	linked genes
genotype	phenotype
gametes	haploid
diploid	sister chromatids
segregation of alleles	independent assortment of alleles

OBJECTIVES

Seldom is the inheritance of characteristics as simple as the transmission of a single pair of dominant/recessive alleles, but understanding this mechanism of inheritance is essential to understanding more complex types. In the exercises that follow, you will learn how to investigate the patterns of inheritance when considering alleles on one pair of homologous chromosomes (**monohybrid cross**); on two pairs of homologous chromosomes (**dihybrid cross**); when genes are located on the X chromosome, a sex chromosome, (**sex-linked inheritance**); and when different genes are on the same chromosome (**linked genes**). You will also see how a **test cross** is used to determine whether an individual is homozygous dominant or heterozygous.

By the end of the exercises you should be able to:

1. Solve and analyze monohybrid, dihybrid non-linked, dihybrid linked, sex-linked, and test crosses.
2. Determine the number of different types of gametes an individual can form, given the individual's genotype.
3. Predict the expected phenotypic and genotypic ratios of monohybrid, dihybrid (linked and non-linked), and sex-linked crosses.
4. Obtain and analyze "observed" results from experimental crosses and interpret the results when compared to the "expected" results.
5. Solve genetic word problems.

MATERIALS

    desktop computer
    display device (e.g., LCD projector)
    Genetics Biology Explorer software, published by Wings for Learning
    small paint brushes
    FlyNap
    foam stopper with FlyNap wand
    Drosophila melanogaster: F_2 of a cross between wild type males and vestigial-winged, ebonybodied, white eyed females
    internet access
    Virtual Fly Lab (Web-based *Drosophila* genetics program)

I. HYPOTHESIS TESTING AND PROBLEM SOLVING USING COMPUTER MODELS

In following exercises, you will be investigating Mendelian inheritance by examining the inheritance of traits among computer generated rabbits, butterflies, and flies. The computer applications simulates the reproduction of diploid organisms which reproduce sexually. The outcome of genetic crosses are based upon the outcome of meiosis and random fertilization. The data generated during these computer-simulated matings will allow you to test your assumptions about how traits are inherited.

Your investigations will employ the scientific method, meaning you will: 1) propose alternate hypotheses for the pattern of inheritance of traits; 2) make predictions based on your hypotheses for the outcomes of particular crosses; 3) test your hypotheses by performing relevant crosses; 4) evaluate your hypotheses by comparing the observed outcome of each cross to the expected for the competing hypotheses.

Your instructor will introduce you to the software application and conduct the breeding simulations based upon your directions. You will work in pairs, and each group will be asked to participate by contributing alternate hypotheses and predictions. Be prepared to write these on the blackboard.

A. Problem #1 - Inheritance of an ear characteristic (straight ears vs. floppy ears) in rabbits

Four rabbits appear on the screen. As you can see, these rabbits exhibit two different phenotypes: straight ears and floppy ears. The two straight-eared rabbits are *true breeding* (*pure breeding*) individuals. This means that if they were crossed they would produce only straight-eared rabbits, and crossing their offspring would also result in only straight-eared rabbits. The two floppy-eared rabbits are also true breeding. What does this tell you about the genotypes of true-breeding individuals?

What do you think will happen when a true-breeding straight-eared rabbit is crossed with a true-breeding floppy-eared rabbit?

Hypothesis #1:
floppy dominate FF vs. ff

Hypothesis #2:
Straight dominate

Your instructor will perform the cross.

Result of computer cross: *Straight dom. (F₁) first generation*

Based on what you know at this point, what are the most likely genotypes of the true breeding parents in this cross? Use "A" to represent the dominant allele and "a" to indicate the recessive allele.

straight-eared: __AA__ ; floppy-eared: __aa__

Can you tell from the results of this first cross, which trait is dominant? Why?

What are the genotypes of the offspring of the first cross (the F₁ generation)? _____
heterozygotes

The principles of Mendelian genetics allow us to predict the expected phenotypes and genotypes of the offspring of this cross, and it is important that you understand the basis of this prediction. The patterns of Mendelian inheritance can be explained based on the events that occur during the production of eggs and sperm (gametes) during **meiosis** and the eventual random reuniting of gametes during **fertilization**. The process of meiosis reduces the number of chromosomes present in gametes to half the number (haploid) present in diploid cells. This reduction in chromosome number happens in a predictable manner.

The following illustrates the production of gametes in the heterozygous parents of the F_1 generation. Some of the chromosomes have been drawn for you, but you need to complete the diagram by drawing in the chromosomes at each step. The gamete types that result from this meiotic process will allow you to predict, using the following Punnett square, what the resultant phenotypes and genotypes will be in a cross between individuals of the F_1 generation. Because the theory of Mendelian inheritance is a "probabilistic" theory you can not only determine what kinds of offspring would result, but also predict the ratios of these types of offspring. Even though the theory predicts the most *likely* (or probable) outcomes of genetic crosses, the actual outcomes may not *exactly* match the predictions—although the results should cluster closely around the prediction if you are analyzing a large enough sample.

These individuals are from the F₁ generation ⟶

Parent 1 X **Parent 2**

To illustrate the events of meiosis, chromosomes can be represented as rod-like structures, each labeled with the allele it carries. The chromosomes of each individual occur in pairs called **homologous chromosomes**. The drawings to the right show each chromosome before it has been duplicated in preparation for meiosis.

Prophase I - each chromosome has been duplicated. The two copies of each chromosome lie side by side and are called **sister chromatids**. The oval outline represents the cell.

Metaphase I - the homologous chromosomes line up next to each other at the center of the cell.

The first cell division separates the homologous chromosomes.

Metaphase II - sister chromatids, still attached to each other, line up in the center of the cell.

The second cell division separates the sister chromatids.

Gametes - each contains one chromosome from each homologous pair, rather than both members of each homologous pair.

These are the gametes you will use to set up the Punnett square that follows.

134   Introduction to Biology: Laboratory Exercises

Complete the Punnett square below using the gamete types you produced from the F_1. Place the gametes of the parents on the outside of the Punnett square and record the predicted genotypes of the offspring (F_2) on the inside.

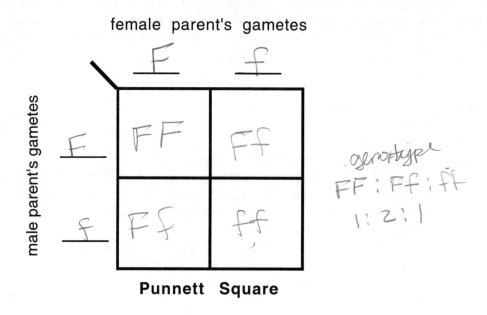

Punnett Square

genotype
FF : Ff : ff
1 : 2 : 1

Based on the genotypes in the Punnett square, what are the expected phenotypes and phenotypic ratio among the offspring of the second cross (the F_2 generation)?

Expected phenotypes ____floppy & straight ear____
Expected phenotypic ratio: ____3:1____

Your instructor will now cross two of the F_1 individuals and produce 40 F_2 offspring. Enter the number of each phenotype below and calculate the observed phenotypic ratio.   out of 30

# of straight-eared: ____24____   239
# of floppy-eared: ____6____   81
Observed ratio: ____4:1____

Should the expected and the observed phenotypic frequencies among the F_2 offspring be exactly the same? Explain.

Your instructor will demonstrate the effects of sample size on the outcome of this probabilistic process.

Genetic Problem Solving Techniques

The most effective way to solve these problems is not to begin by randomly mating individuals, but by formulating specific hypotheses concerning the genotypes of the parents, and determining the predicted offspring for each hypothesis. Next, perform the necessary crosses and analyze the resulting data to support or reject your hypotheses. In other words, you need to generate all the reasonable possibilities for what the genotypes of the parents might be (these are your hypotheses), and then test these hypotheses by producing offspring and determining which of the hypotheses best explains the results.

For each of the problems, follow this procedure:

1. Formulate your hypotheses
2. Make your predictions before the crosses are performed.
3. Use what you know about characters determined by a single gene with two alleles to generate several alternative possibilities for the genotypes of the unknown individuals.
4. Use Punnett squares to predict what phenotypic frequencies are expected among the offspring for each alternative hypothesis.
5. Perform crosses among the unknown individuals and compare the observed results with the expected results to eliminate some of the initial possibilities. Conduct additional crosses to narrow the alternatives down to the best explanation.

B. Problem #2 - Inheritance of body color (black body vs. white body) in butterflies

Three butterflies appear on the screen; two black-bodied males and a white-bodied female. As you can see, these butterflies exhibit two different phenotypes: black body and white body. You do not know if any of the individuals breed true.

Hypothesis #1:

White body color is caused by a dominant allele. List all the possible genotypes for each parent below:

Complete Punnett squares and determine the expected phenotypic ratios among the offspring for crosses among the genotypes you listed above.

	W	w̄
w̄	Ww̄	w̄w̄
w̄	Ww̄	w̄w̄

	W	W
w̄	Ww̄	Ww̄
w̄	Ww̄	Ww̄

Expected phenotypic ratio: __1:1__   Expected phenotypic ratio: __100% W__

Hypothesis #2:
Black body color is caused by a dominant allele. List all the possible genotypes for each parent below:

Complete the Punnett squares below, and determine the expected phenotypic ratios among the offspring for crosses among the genotypes you listed above.

	b	b
B	Bb	Bb
b	bb	bb

Expected phenotypic ratio: __1:1__

	b	b
B	Bb	Bb
B	Bb	Bb

Expected phenotypic ratio: __100%__

Your instructor will now perform the crosses. Record below the results of each cross performed. Be sure to record the results of the crosses of both individual males with the female.

Black body male #1 crossed to white body female: b47w53 male

Black body male #2 crossed to white body female: Black dominant

Which of your hypotheses best matches the observed results?

Final conclusions:

Parental genotypes:
Black body male #1 - __heterozygotic__
Black body male #2 - __homozygotic dominant__
White body female - __homozygotic recessive__
Dominance relationship among alleles: _____

What can you conclude about crosses which involve a homozygous recessive individual and a phenotypically dominant individual?

What is the general name for this type of cross?

Exercise 8 — Genetics 137

C. Problem #3 - Inheritance of fur color in rabbits

Three rabbits appear on the screen; two dark-brown males and a white female. As you can see, these rabbits exhibit two different phenotypes: dark-brown fur and white fur. You do not know if any of the individuals breed true.

Hypothesis #1:
Dark-brown fur is caused by a dominant allele. List below all the possible genotypes for each parent.

M - Dd, DD
F - dd

Draw Punnett squares for each possible cross you could perform and fill them in to determine the expected phenotypic ratios among the offspring for crosses among these genotypes.

Hypothesis #2:
White fur is caused by a dominant allele. List below all the possible genotypes for each parent.

♂ M = ww
♀ F = Ww, WW

Draw Punnett squares for each possible cross you could perform and fill them in to determine the expected phenotypic ratios among the offspring for crosses among these genotypes.

Your instructor will now perform the crosses as directed by selected groups. What are the results of the crosses and your conclusions based on these crosses?

[handwritten: light brown]
[handwritten Punnett square with E^P, E^D across top and E^d, E^d on side, producing E^P E^d, E^D E^d offspring]
[handwritten: incomplete dominance]

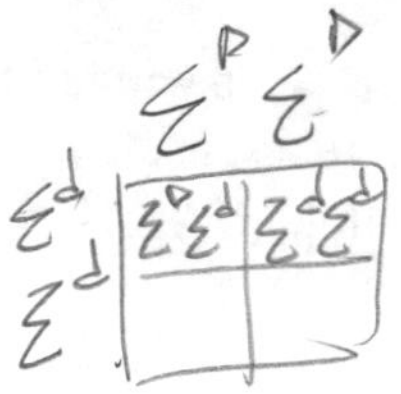

What possibility was overlooked when you predicted the possible genotypes of the parents and the resultant phenotypes of their offspring? *[handwritten: incomplete dominance]*

What modification(s) do you need to make to your original hypotheses? That is, are there additional hypotheses you need to test?

[handwritten: mate 2 of light brown]

Additional hypotheses:

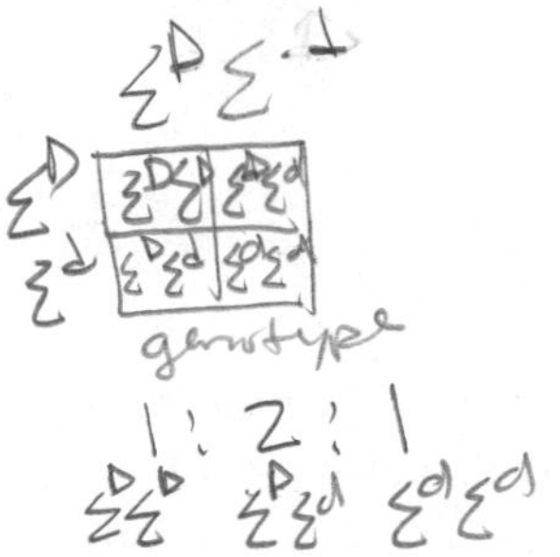

[handwritten Punnett square with E^P, E^d across top and E^D, E^d on side]
[handwritten: genotype 1:2:1, E^D E^D, E^D E^d, E^d E^d]

Your instructor will now perform the additional crosses you suggest to test your hypotheses. Generate Punnett squares which describe these hypotheses, and the results of additional crosses that test these hypotheses.

Final conclusions:

    Parental genotypes:
        Dark-brown male #1 - _____
        Dark-brown male #2 - _____
        White female - _____
    Dominance relationship among alleles: _____

Exercise 8 — Genetics

D. Problem #4 - Inheritance of eye color (red vs. white) and wing color (clear vs. gray) in flies

Two flies appear on the screen, a white-eyed/gray-winged male and a red-eyed/clear-winged female. This problem deals with the inheritance of two characters determined by two genes, each with two alleles. The genes may be assumed to be on separate chromosomes, i.e., non-linked. Again, you do not know if any of the individuals breed true.

Before you begin to tackle this problem, think about the approach you should use and the questions you wish to answer first. For example, what are the hypotheses concerning the dominance relationships among the alleles, and what crosses could you perform to generate data to support or reject these hypotheses? Should your group try to solve the problem simultaneously for both characters, or is there another approach you could take? If you try and solve for both characters simultaneously, how many hypotheses are there? List them.

It should be obvious that considering both characters simultaneously complicates the problem and results in too many hypotheses to test. A better approach is to understand the mechanism of inheritance for each character before you try to formulate hypotheses that involve simultaneous consideration of both characters.

Inheritance of eye color:

Is white eyes or red eyes dominant? What are the four possible hypotheses? List them.

Red eyes & clear wings
White eyes & gray wings
red " & gray
white & clear

♂ WW, Ww
♀ ww

♀: RR, Rr
♂: rr

Set up a Punnett squares to generate the predicted genotypes and phenotypes of the offspring from each hypothesis. When you have completed your Punnett squares, your instructor will perform the crosses the class has suggested.

Punnett squares:

What can you conclude about the dominance relationship in eye color?

red dominat

What are the genotypes of the parents?

red-eyed female: *RR*

white-eyed male: *rr*

What are the genotypes of the F_1 from the supported hypothesis:

female: *Rr*

male: *Rr*

If you cross the F_1 offspring, what are the expected genotypes and phenotypes in the F_2 with respect to eye color? What is the expected phenotypic ratio? Draw a Punnett square of this cross.

red : write, 3:1

Punnett square:

F_2 Phenotypes: _____

Expected phenotypic ratio: *3:1*

Your instructor will now perform the cross among the F_1 offspring to produce 100 offspring. List the phenotypes and the number of each below:

What is the observed phenotypic ratio? _____ Is the observed phenotypic ratio close to the expected ratio?

Inheritance of wing color:

Is gray or clear wing dominant? What are they possible hypotheses? List them.

clear ♂ M - c̄c̄
♀ F - CC, Cc̄
gray ♂: GG, Gg
♀ = gg

Set up a Punnett squares to generate the predicted genotypes and phenotypes of the offspring for each hypothesis. When you have completed drawing your Punnett squares, your instructor will perform the crosses you have suggested.

Punnett squares:

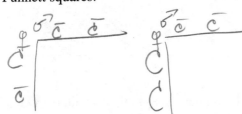

clear

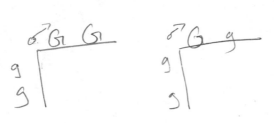

gray

What can you conclude about the dominance relationship in wing color?

clear

What are the genotypes of the parents?

clear-winged female: __CC__

gray-winged male: __c̄c̄__

What are the genotypes of the F₁ from the supported hypothesis:

female: __Cc̄__

male: __Cc̄__

If you cross the F_1 offspring, what are the expected genotypes and phenotypes in the F_2 with respect to wing color? What is the expected phenotypic ratio? Draw a Punnett square of this cross.

Punnett square:

F_2 Phenotypes: _CC, Cc, Cc, cc_

Expected phenotypic ratio: _3:1_

Your instructor will now perform the cross among the F_1 offspring to produce 100 offspring. List the phenotypes and the number of each below:

What is the observed phenotypic ratio? _____ Is the observed phenotypic ratio close to the expected ratio?

Considering both eye and wing color simultaneously:

At the beginning of this problem you were told to assume that the genes for wing and eye color were on separate pairs of homologous chromosomes, i.e., non-linked. Let's explore the implications of these genes being non-linked.

If two genes are non-linked, the alleles of each gene assort independently of each other in gametes during meiosis. If we were to perform a cross between two F_1 individuals, we would produce the F_2. Given your knowledge of the parental (P_1) genotypes for eye and wing color that you discovered above, by analyzing each gene separately, and the expectation of independent assortment, you can predict the phenotypes and their frequencies in the F_2. How?

First, what proportion of the F_2 flies would be expected to have red eyes? _75%_ White eyes? _25%_ Consult the Punnett square you created above for the F_2 when you investigated the inheritance of eye color. Next, what proportion of F_2 flies would be expected to have clear wings? _75%_ Gray wings? _25%_ Again, consult the Punnett square you generated for your investigation of wing color.

If assortment of the chromosomes and their genes is independent, then whether an individuals is red or white-eyed has no bearing on whether it's clear or gray-winged. The probability of two independent events both occurring together is the product of their separate probabilities. (For example, if you flip a quarter and a silver dollar, the probability that both will be heads is 1/2 X 1/2 = 1/4). Thus, the proportion of F_2 expected to have both red eyes and clear wings is the product of the probability of having red eyes times the probability of having clear wings. Fill out the table below for each of the phenotypes.

Eye color		Wing color		Product of probability	Percent
phenotype	prob.	phenotype	prob.		
red	3/4	Clear	3/4	9/16	56%
red	3/4	Gray	1/4	3/16	16.75%
white	1/4	Clear	3/4	3/16	16.75%
white	1/4	Gray	1/4	1/16	6.25%

Another way to predict the results of this cross is with a Punnett square. Complete the Punnett square below for a cross between two F_1 individuals.

female gametes

	RC	Rc̄	rC	rc̄
RC	RRCC	RRCc̄	RrCC	RrCc̄
Rc̄	RRCc̄	RRc̄c̄	RrCc̄	Rrc̄c̄
rC	RrCC	RrCc̄	rrCC	rrCc̄
rc̄	RrCc̄	Rrc̄c̄	rrCc̄	rrc̄c̄

(male gametes)

Predicted phenotypic ratio: 9:3:3:1

RC

Your instructor will now conduct the F₁ X F₁ cross. The sample size will be set to 100 individuals so it will be easy to convert the results to percentages. Record your results below:

Number of:

red-eyed/clear-winged ____9____

red-eyed/gray-winged _____

white-eyed/clear-winged ____19____

white-eyed/gray-winged ____1____

Calculate the observed phenotypic ratio: _____

Are the predicted and observed phenotypic ratios the same? _____ Why?

E. Problem #5 - Inheritance of eye color (red vs. white) and wing color (clear vs. gray) in flies, but with the genes linked to the same chromosome

Now consider the same cross as in the previous problem: white-eyed/gray-winged male X redeyed/clear-winged female. However, now instead of the genes for eye and wing color being on different homologous chromosomes, we have used the software application to "genetically engineer" them to be located very close together on the same pair of homologous chromosomes. That is, we have created a situation were the genes are closely **linked**. Because they are so close together, no crossing-over occurs during meiosis. All other aspects of the genetics of this cross are the same as those in Problem #4, i.e., the dominance relationships of the alleles are the same, and both of the parents are homozygous (female = AABB and male = aabb).

In the space below, set up a Punnett square for a cross of the P₁. Remember, the genes are now linked to the same pair of homologous chromosomes and this will have a bearing on how many different gamete types are produced because assortment of the alleles of the two genes is no longer independent.

♀ AABB
♂ aabb

Punnett square:

	AB
ab	AaBb

Chromosome ⎰A
⎱B
↓

What F₁ phenotypes do you expect? __Red & Clear__

What is the expected F₁ phenotypic ratio? _____

Your instructor will perform the cross described above and produce 100 offspring. Are the results consistent with your predictions?

A cross of the F₁:

Because the genes are linked to the same pair of homologous chromosomes the possible kinds of gametes is different than if the genes were not linked. The key to solving this problem is generating the correct kinds of gametes from each parent. In the space below, draw the chromosomes of two F₁ parents in a cross to obtain the F₂, and indicate the location of the genes on the chromosomes.

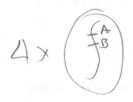

What types of gametes can the parents produce? __AB, ab__

Next, complete a Punnett square of a cross between the F₁.

Punnett square:

	AB	ab
AB	AABB	AaBb
ab	AaBb	aabb

What F₂ phenotypes do you expect? __Red : Clear, White : Gray__

What is the expected F₂ phenotypic ratio? __3:1__

Your instructor will perform the cross described above and generate 100 offspring. Are the results consistent with your predictions?

The results of a dihybrid cross involving linked genes looks like what kind of cross?
__monohybrid cross__

146 Introduction to Biology: Laboratory Exercises

II. ANALYZING A GENETIC CROSS IN *Drosophila melanogaster*

Now that you have investigated inheritance using computer simulations, its time for you to analyze a genetic cross involving a real organism, *Drosophila melanogaster*, a species of fruit fly. The fruit fly is used extensively in genetic research. It is small, which makes it easy to handle, and inexpensive. Two of the best attributes of this organism as a genetic research tool are its many distinguishing mutant characteristics, and its short life cycle. The latter allows the researcher to obtain many generations in a short period of time.

Your instructor will project digital photographs illustrating the differences among males and females, as well as photos illustrating both the normal and mutant traits you will be investigating.

Sexing and Recognizing Mutant Traits of *Drosophila*

A prerequisite for working with *Drosophila* is the ability to distinguish the sexes and to recognize the mutant characteristics under consideration. Mutant traits are those that differ from the normal traits in the population. For example, if a survey were made of a wild population of fruit flies, the predominant eye color would be red. Occasionally an individual with white eyes might be found in a population. White eye would be a **mutant** eye color in the population, while red eye color would be the normal or "wild type."

Sexing

Two general characteristics which are used to sex fruit flies are body size and pigmentation of the abdomen. Males are generally smaller than females and have a dark-tipped abdomen. Both characteristics may be confusing in immature flies and should never be the sole criteria on which a determination of sex is made. The best characteristic for determining sex is the presence of a **sex comb** on either of the forelegs of the male (Figure 8-4). No such structure is present on females.

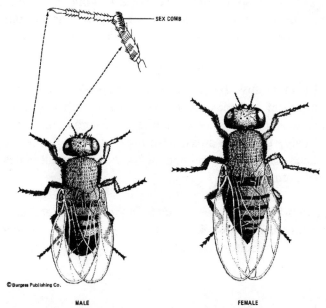

FIGURE 8-4. Male and Female *Drosophila*

Exercise 8 — Genetics 147

Recognizing Mutant Traits

The "wild type" eye color is red, and the "wild type" shape of the wings should appear similar to those in Figure 8-4. The mutants are white-eyed and vestigial-winged, respectively. The white-eyed trait should be self-explanatory. Vestigial wings are reduced in size and malformed. Normal or "wild type" pigmentation of the body is more difficult to describe. The best way to become familiar with normal pigmentation is to look at a large number of wild type flies, and compare the degree of pigmentation to the mutant trait, ebony body. Ebony-bodied flies have a more darkly pigmented body.

A word of caution: newly emerged flies may not appear dark. The population you are analyzing should contain both darkly pigmented and pale-appearing flies. If you think your sample contains only pale-appearing flies, consult your lab instructor for assistance and directions. Also, newly emerged flies may have wings that look similar to vestigial wings.

Handling Fruit Flies

To save time in the lab, your instructor will describe the techniques used in handling fruit flies. The flies are anesthetized with FlyNap, which is extremely volatile and flammable. In addition, FlyNap can irritate eyes and skin, so use with caution. No open flames or cigarettes will be allowed in the lab.

The Experimental Cross

In the mitosis/meiosis lab you were intentionally led to believe, for simplification, that all sexually reproducing organisms had a certain number of homologous chromosomes and no other kinds. This is not strictly the case. For example, in *Drosophila* and humans, the expression of sex is controlled by one pair of chromosomes called the **sex chromosomes**. These chromosomes are arbitrarily called X and Y. Females have the sex chromosome complement of XX, while males are XY. The sex chromosomes are not considered homologous because the X and Y chromosomes do not have identical linear gene arrangements along their lengths. Any chromosome that is not a sex chromosome is called an **autosome**.

Approximately five weeks before this lab period, a cross was made between male flies that were normal for eye color, wing shape, and body color, and female flies that were abnormal or mutant for these characteristics. This was the P_1 cross. A week after the P_1 cross, the parent flies were removed and discarded. Two weeks after the initial cross, offspring (F_1) of the P_1 flies began to emerge from pupal cases. All F_1 flies had normal wings and normal body color, but all males were white eyed while all females were red eyed.

Ten female and ten male flies from the F_1 were transferred into bottles containing new growth medium and allowed to reproduce. One week after this transfer, the F_1 flies were removed and discarded. The vials of flies provided for you contain the F_2, offspring of the F_1.

The following is a summary of the cross between normal males and mutant females up to and including the F_1 cross.

	males		females
P_1	Normal-winged, normal-bodied, red-eyed	X	Vestigial-winged, ebony-bodied, white-eyed

↓

F_1 Normal-winged, normal-bodied, red-eyed (females)
Normal-winged, normal-bodied, white-eyed (males)

Cross of F_1:

	Normal-winged, normal-bodied, white-eyed	X	Normal-winged, normal-bodied, red-eyed

↓

F_2 ?

Use the following convention for labeling your genetic crosses:

    V = normal wing
    v = vestigial wing
    E = normal body color
    e = ebony body color
    X^R = red eye color
    X^r = white eye color

Even though this cross involves three separate traits, it is possible (and much preferable) to consider one trait at a time, just as you did in D Problem 4. For example, to observe how wing shape is inherited you should ignore eye and body color, treating these two traits as if all the flies in the F_2 are normal for them. This will allow you to view the results as if the experiment were a simple monohybrid cross with respect to wing shape. This same procedure can be used to view body color and eye color as monohybrid crosses. You will also consider body color and wing shape as a dihybrid cross (in this case you should ignore eye color).

Exercise 8 — Genetics

General Instructions

Since no further crosses will be made, you should over-anesthetize the entire population of F_2 flies. Over-anesthetizing will kill the flies and make handling them easier. Your instructor will demonstrate how to do this. You will use the same flies to obtain results for each type of cross.

A. Monohybrid Cross with Respect to Wing Shape

Procedure

Ignore eye color and body color and count the number of normal-winged flies and the number of vestigial-winged flies in the F_2. Insert these numbers in the appropriate spaces below and calculate a phenotypic ratio.

	males		females
	NN ~~Nn~~		nn
P_1	Normal-winged	X	Vestigial-winged
F_1		Normal-winged (all flies)	

Cross of F_1

	Normal-winged	X	Normal-winged
F_2		?	

F_2 Analysis:

    # of normal-winged flies   183   8:1

    # of vestigial-winged flies   23

    Observed phenotypic ratio: 8:1
                                                     N   V

Which trait is dominant?   normal

What are the genotypes of the P_1?   NN & nn

What is the genotype of the F_1?   Nn

What is the expected genotypic ratio of the F$_2$: __1:2:1__

What is the expected phenotypic ratio F$_2$: __3:1__

Are the observed and expected phenotypic ratios of the F$_2$ exactly the same? __No__ Why or why not? __natural selection__

B. The Test Cross

If you have a normal-winged fly, how can you tell whether the fly is homozygous or heterozygous for the normal-winged allele? To determine the genotype of a phenotypically dominant individual you need to carry out what is called a **test cross**. A test cross involves crossing a phenotypically dominant individual with a homozygous recessive individual. Depending on which of two outcomes you obtain from such a cross, you will be able to ascertain whether an individual is homozygous dominant or heterozygous. For example, consider the following cross, in which a normal-winged fly was crossed with a vestigial-winged fly, and all of the progeny were normal-winged.

P$_1$ V _?_ (normal-winged)   X   vv (vestigial-winged)

F$_1$ All normal-winged

In this case is the dominant parent homozygous or heterozygous? __homozygous__

Next, consider the alternative outcome which is illustrated below:

P$_1$ V _?_ (normal-winged)   X   vv (vestigial-winged)

F$_1$ 1/2 normal-winged
        1/2 vestigial-winged

From these results, was the normal-winged parent from this cross homozygous or heterozygous? __heterozygous__. You should note that all the genetic crosses you will be studying begin with a test cross. Why?

C. Monohybrid Cross with Respect to Body Color

Procedure

Ignore all traits except body color, and count the number of normal-pigmented flies and the number of ebony-bodied flies. To ensure that you are correctly identifying body color, you may want isolate a normal-pigmented fly and an ebony-bodied fly to use for comparison as you segregate your files for counting. Enter the phenotypic results in spaces provided.

	males		females
P_1	Normal-pigmented NN	X	Ebony-bodied nn

↓

F_1 Normal-pigmented (all flies)

Nn (males)   Nn (females)

Cross of F_1:

Normal-pigmented   X   Normal-pigmented

↓

F_2 ?

F_2 Analysis:

of normal-pigmented flies __147__

# of ebony-bodied flies __14__

Observed phenotypic ratio: __15:1__

Which trait is dominant? __normal__

What are the genotypes of the P_1? _____

What is the genotype of the F_1? _____

What is the expected genotypic ratio of the F_2: __1:2:1__

What is the expected phenotypic ratio F_2: __3:1__

Are the observed and expected phenotypic ratios of the F₂ exactly the same? _____ Why or why not?

Would you expect ebony-colored flies in the F₂ to breed true if crossed with each other? That is, would the offspring always be ebony colored? _____ Why?

D. Dihybrid Cross with Respect to Body Color and Wing Shape

Are the genes controlling body color and wing shape linked to the same chromosome or are they found on separate homologous chromosomes? This is the central question of the exercise.

Procedure

Disregard eye color and count the number of each phenotype in the sample with respect to body color and wing shape. Record the number and kind of each phenotype in the spaces provided below.

P₁   Normal-winged, normal-bodied   X   Vestigial-wing, ebony-bodied

VvEe   *VvEe*

F₁   Normal-winged, normal-bodied (all flies)

Cross of F₁:

Normal-winged, normal-bodied   X   Normal-winged, normal-bodied

F₂   ?

F₂ Analysis:

   Phenotypes and their numbers:

Observed phenotypic ratio: 38:7:6:1

What are the genotypes of the P₁? _____

What is the genotype of the F₁? _____

In the space below, set up a Punnett square for a cross of the F₁.

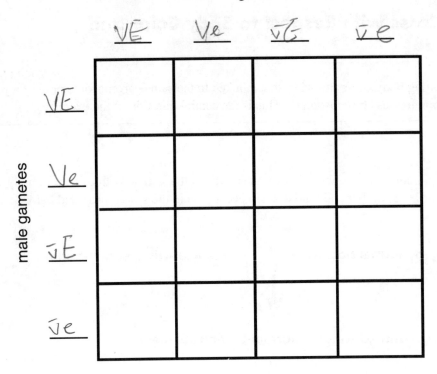

What is the expected F₂ phenotypic ratio: 9:3:3:1

Are the genes for body color and wing shape linked or non-linked? _____ How do you know?

E. Inheritance of Eye-color

Procedure

In *Drosophila* the gene for eye-color is sex-linked, i.e., the gene for eye color is found on the sex chromosomes. More specifically, the gene for eye color is linked to the X chromosome. No gene for eye color is found on the Y chromosome. Therefore, females fruit flies have two X chromosomes, and can either be homozygous or heterozygous. Males on the other hand have only one X chromosome, and the gene carried on the X will always be expressed in their phenotype.

The convention for labeling alleles of the sex chromosomes is to show them as superscripts on the X chromosome. Thus, if alleles A and a were sex-linked, we would designate them as X^A and X^a. A female heterozygous for the alleles would be $X^A X^a$, while a male can only be $X^a Y$ or $X^A Y$.

The following outlines the crosses that were made to obtain the F_2. Separate the F_2 flies according to sex, and count the number of red-eyed and white-eyed flies of each sex. Enter these numbers in the space provided below.

P₁ Males: red-eyed X Females: white-eyed

F₁ Males: white-eyed (all)
Females: red-eyed (all)

Cross of F₁ white-eyed X red-eyed

F₂ Phenotypes:

  males: Number of red-eyed _____

  Number of white-eyed _____

  females: Number of red-eyed _____

  Number of white-eyed _____

Observed phenotypic ratio in

  males: _____

  females: _____

In the space below draw a Punnett square illustrating the cross of the F_1 to obtain the F_2 progeny.

Exercise 8 — Genetics   155

How many different types of gametes can the males of the F₁ produce? The females? List them.

    Males: _____   Females: _____

What are the possible genotypes, by sex, of the F₂?

        Males: Females:

What if the eye color of the parents were reversed? Would the results of the cross be the same? If the males of the P₁ are white-eyed and the females are homozygous dominant for red eye color, what is the pattern of inheritance of eye color through the F₂? In the space provided below fill in the phenotypes and genotypes of the F₁ and F₂.

           Males   **Females**

P₁   white-eyed   X   red-eyed

⬇

F₁ Phenotypes: **F₁ Genotypes:**

    males: _____   **males:** _____

    females: _____   **females:** _____

Cross of F₁   X

⬇

F₂ Phenotypes: **F₂ Genotypes:**

    males: _____   **males:** _____

    females: _____   **females:** _____

III. VIRTUAL FLY LAB

If time permits, your instructor will demonstrate a World Wide Web (WWW) application called Virtual Fly Lab, located at the University of California, Los Angles . This site allows you to simulate the crossing of two fruit flies to determine the pattern of inheritance of nine phenotypic traits (e.g., bristles, eye color, etc.). You can simulate the actual fly cross you just analyzed.

Specifically, you can cross a true-breeding red-eyed/normal-winged/normal body color male with a true-breeding white-eyed/vestigial-winged/ebony-bodied female to produce offspring. You can then cross those offspring to produce an F_2 generation.

You can find Virtual Fly through the "Web Library" of WWW links at the General Biology Program Web site. Virtual Fly can be found within the Mendelian Genetics section of the library (http://genbiol.cbs.umn.edu/gbweb.html#mend), or you can open the following URL location with your Web browser: (http://vflylab.calstatela.edu/edesktop/VirtApps/VflyLab/Design.html). You can get access to the WWW from the General Biology Program tutorial room, P176 Kolthoff Hall, or from one of the many public computer labs on campus.

Plant

cross pollination → new plant varieties, better survival
agents - wind dispersal, water, animals

pollination agent (if animal, morphology & behavior of animal
determines which plant to pollinate)

provides - on bees & butterflies

<u>primary function of fruit is seed dispersal</u>

II.
1) Roots - anchors, water/nutrient procurement, food storage } veg
2) Stems - support leaves
3) Leaves - photosynthetic site
4) Flowers - reproductive organ } rep.
5) Fruits - reproductive organ, contain seed.

9 Plant Biology

INTRODUCTION

Members of the Kingdom Plantae comprise approximately 21 percent of all known living species. The Kingdom contains approximately 264,000 species, of which, most (89%) belong to a group called the angiosperms (flowering plants). Though the plant kingdom does not constitute the majority of species, its role in the biosphere is crucial to the survival of the vast majority of life forms. The Kingdom Animalia, with a single phylum Arthropoda containing approximately one million species, holds that distinction.

According to one common classification scheme, the Kingdom Plantae contains ten divisions (=Phyla). From Table 9-1 you can see that the flowering plants are by far the most numerous. Therefore, most of the exercises that follow will concentrate on the characteristics of this important group.

TABLE 9-1. Classification Scheme for the Kingdom Plantae

Division	Common Name	Number of Species
Bryophyta	Mosses,	10,000
Hepatophyta	Liverworts	6,500
Anthocerophtya	Hornworts	100
Psilophyta	Whiskferns	10-13
Lycophyta	Club mosses	1,000
Sphenophyta	Horsetails	15
Pterophyta	Ferns	12,000
Coniferophyta	Conifers	550
Cycadophyta	Cycads	100
Ginkgophyta	Ginkgos	1
Gnetophyta	Gnetophytes	70
Anthophyta	Flowering plants (angiosperms)	235,000

Even the most complex individuals in the plant kingdom are much simpler in their structure than most multicellular animals. For instance, most multicellular animals are made up of several highly organized and specialized organs, such as the heart, lungs, kidneys, etc. In turn, the organs of the body are organized into "organ systems," such as the circulatory system that includes the heart, blood vessels, and the lymphatics. Flowering plants on the other hand contain relatively few organs: roots, stems, leaves, flowers, and fruits. The number and placement of these organs are also highly variable in plants in contrast to the animal kingdom, where there has been relative constancy in the number of organs through phylogenetic lines. In addition, the different organs of the plant are not tied together by a nervous system, and often act as separate units.

An advantage to this relatively loose organization is that it helps plants adapt to the environments in which they grow. Unlike animals, which are mobile and can seek out suitable places to live and grow, plants are sessile. Once rooted in a location, a plant must respond to the local environment in appropriate ways or perish. By controlling the number, size, and placement of its organs, a plant can overcome some disadvantages of being stationary.

The sedentary state of plants also makes them vulnerable to predation by a wide host of animals. It is astonishing that plants can survive the relentless onslaught of so many herbivores (a heterotrophic animal that eats plants). Yet upon close examination we discover a line of defensive adaptations (chemical toxins, stinging agents, impaling structures, etc.) that should be the envy of the Pentagon. A person could devote a lifetime to studying the mechanisms of defense in just one small group of plants.

Another consequence of the immobility of adult plants is their exploitation of animals for transportation. In general, the structures of flowers often induce animals to move pollen between flowers, while the characteristics of fruits often entice animals to disperse seeds. The adaptive strategies of plants for pollination and seed dispersal are truly varied and amazing.

The most obvious and important distinction, which can be made between plants and animals, is that plants have the capacity to harness the energy of sunlight through the process of photosynthesis. This capture of energy is no trivial process, for it is the major avenue through which energy becomes available to most life forms, animals as well as plants. Even though the plant kingdom does not comprise the lion's share of species, plants represent the vast majority of biomass (total weight of organisms) on the planet. This biomass is literally the "food" that supports most life processes.

Plants also differ from animals in having pronounced and conspicuous diploid and haploid multicellular phases, called the sporophyte and gametophyte, respectively, in their life cycles. This phenomenon is generally called **Alternation of Generations**. The haploid and diploid generations are separated from each other by the processes of meiosis and fertilization, with meiosis marking the end of the diploid phase, and with fertilization marking the end of the haploid state. Both the haploid and diploid multicellular phases of plants are the products of mitotic cell divisions. In contrast with animals, the gametes of plants (egg and sperm) are the products of mitosis, not meiosis. Plants vary widely in the specific details of the alternation of generations in their life cycles, but all follow a generalized scheme. In the first section that follows, you will learn this generalized scheme, and then how the details of the alternation of generations vary in three major plant groups, the mosses, ferns, and flowering plants. In subsequent sections you will investigate the general organization and features of the organs of flowering plants.

OBJECTIVES

1. Understand the main features of a generalized life cycle of a multicellular plant.
2. Understand the differences between moss, fern, and flowering plant life cycles.

3. Be able to define the terms gametophyte and sporophyte, and then be able to identify these phases in the life cycles of mosses, ferns, and flowering plants.
4. Be able to describe where the mitotic and meiotic cell division processes occur in the life cycles of mosses, ferns, and flowering plants.
5. Be able to recognize and describe the major functions of roots, stems, leaves, flowers, and fruits.
6. Be able to name at least two root types and give an example of each.
7. Describe the function of the Casparian strip in the root cells of the endodermis.
8. Be able to diagram and label the major structures of a woody twig.
9. Be able to diagram and label a cross section of a typical leaf.
10. Be able to explain the functions of stomata and guard cells in the epidermal layers of leaves.
11. Understand and be able to describe how the morphology of flower parts in different plant species promotes pollination by insects, birds, or wind.
12. Be able to recognize and describe the functions of the four major floral parts: sepals, petals, stamens, and pistils.
13. Be able to diagram and label the major structures of a typical flower.
14. Be able to define and correctly use the following terms: perfect and unisex flowers; monoecious, dioecious, and hermaphroditic plants.
15. Be able to define the terms: seed and fruit.
16. Understand how the various structures of a fruit promote dispersal of seeds by birds, mammals, wind, or other dispersal agents.

MATERIALS

adult fern
moss—gametophytes with sporophytes
adult flowering plants
— perfect flower
— unisex flowers
flowering plants illustrating different pollinator types
lily flowers
fruit types (a variety of species)
cross section of a dicot root
cross section of a dicot leaf
cross section of wood stems (a variety of species)
woody twigs
posters of plant life cycles
dissected flowers of lily and amaryllis (demo)
flower model (demo)
dissecting microscopes
razors
tweezers

I. LIFE CYCLES IN THE PLANT KINGDOM

A. Generalized Life Cycle of a Sexually Reproducing Plant, Figures 9-1 & 9-2

In contrast with the life cycles of animals, the life cycles of plants are characterized by having a distinct alternation of generations. That is, there are both **multicellular** haploid and diploid phases present during some stage of the life cycle. In animals, the multicellular phase is restricted to the diploid state, Figure 9-1.

If you compare the life cycles of plants and animals, Figures 9-1 and 9-2, you will note that, starting at the process of fertilization up to the process of meiosis, the two life cycles are qualitatively identical. At the point of meiosis, the two life cycles begin to diverge. Notice that meiosis produces haploid egg and sperm in the animal cycle, and that the haploid phase is restricted to these unicellular gametes. In contrast, the process of meiosis in plants produces haploid spores, which germinate and grow by mitotic cell divisions into haploid multicellular structures. Eventually, certain cells in the haploid multicellular structure undergo mitotic cell divisions to produce egg and sperm, not more **somatic** (body) cells.

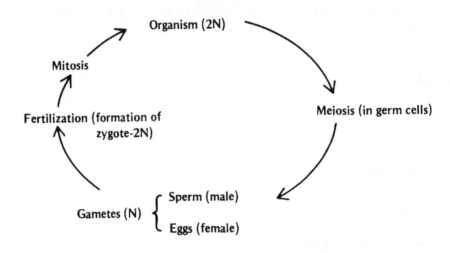

FIGURE 9-1. Generalized Life Cycle of a Sexually Reproducing Multicellular Animal

You should note that the process of mitotic cell division is responsible for the growth of both the diploid and haploid multicellular phases in the life cycle of plants. It is also the process that produces the gametes. Meiosis, on the other hand, is the process that reduces the chromosome number from diploid to haploid, and produces haploid structures called spores. In plants, the diploid multicellular spore producing structures are called **sporophytes**, while the haploid multicellular gamete producing structures are referred to as **gametophytes**.

Before we move on to the general life cycles of some of the major plant groups, I'd like to leave you with the following thought. Suppose humans had a life cycle that was similar to the generalized life cycle in Figure 9-2, would we consider the multicellular haploid structures individuals?

1. What process ends the sporophyte generation? Explain.

   meiosis

2. Name the process that starts the gametophyte generation.

   mitosis

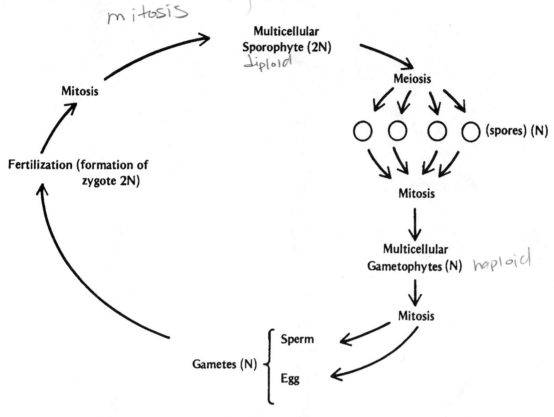

FIGURE 9-2. Life Cycle of a Sexually Reproducing Multicellular Plant

B. Life Cycle of a Moss, Figure 9-3

Mosses lack vascular tissue for transport and support, and are therefore generally small and live in moist environments. Unlike flowering plants, the mosses have a haploid gametophyte that is larger than the diploid sporophyte. The sporophyte is not only smaller than the gametophyte, but is dependent upon the gametophyte for nutrients and support. Mosses are also dependent upon external water for fertilization.

A moss life cycle may begin with the germination of a haploid spore. Through mitotic cell divisions the haploid tissue grows into a spreading network of filaments called a **protonema**. "Leafy" gametophytes grow vertically from the protonema network. The protonema/gametophyte structure is anchored to the substrate by unicellular projections of the protonema called **rhizoids**. The gametophytes that grow from the protonema are of two kinds, those that have sperm producing structures (**antheridia**) or those that contain egg producing structures (**archegonia**). These structures are located at the tops of the gametophytes.

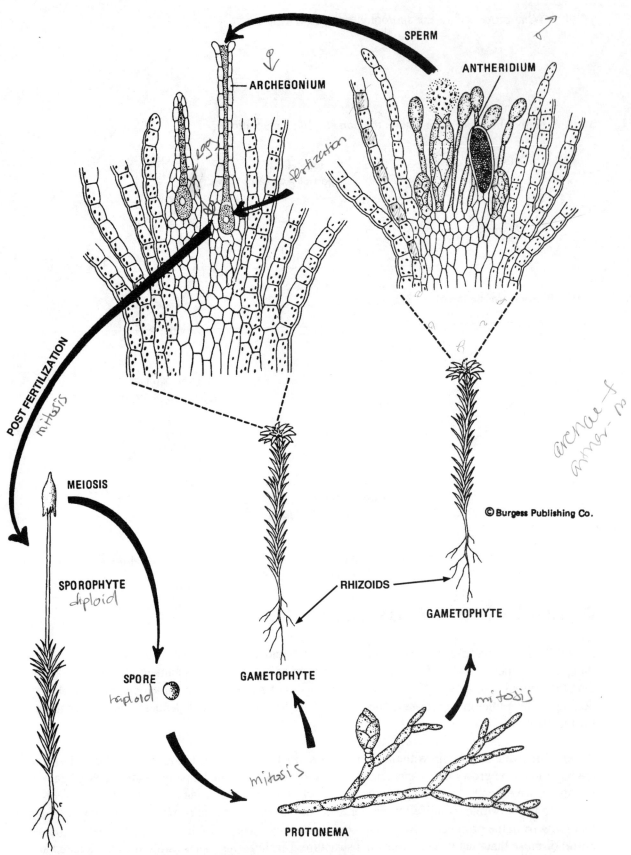

FIGURE 9-3. Life Cycle of a Moss

When an antheridum has matured, i.e., produced sperm, it ruptures and liberates its sperm to the surrounding environment. Often antheridia are ruptured by the impact of raindrops. The splashing water then transports the sperm onto adjacent archegonia, where the sperm can swim down the necks of the archegonia to fertilize the resident eggs.

Once fertilization has occurred, the diploid zygote grows through mitotic cell divisions, into a multi-cellular sporophyte that is anchored in the top of the gametophyte. The sporophyte consists of an elongated stalk that terminates with a large capsule (**sporangium**). The tissues (2N) inside this capsule undergo meiosis to produce a multitude of haploid spores. When the spores are mature, the capsule of the sporophyte ruptures and the spores are discharged to start the cycle anew.

1. Which generation of the moss life cycle is considered the dominant generation? Explain.

    Gametophyte, depends on it for nutrients & support.

2. Would you expect moss species to be abundant in dry habitats? Explain.

    No, they depend on water to reproduce. & don't have vascular tissue

C. Life Cycle of a Fern, Figure 9-4

Ferns are vascular plants, that is, they contain specialized tissues for conducting water, nutrients, and photosynthetic products throughout the plant body. They have well-developed leaves (**fronds**), but their stems usually consist of underground creeping structures called **rhizomes**. The spore producing structures (**sporangia**) are often found on the underside of fronds or, sometimes are found on nonphotosynthetic leaves that are specialized for reproduction. Most species of ferns are tropical, but because ferns have true roots and vascular tissue, they are able to exploit environments that are typically dryer than the environments in which we find mosses. Even though ferns can live in relatively dry environments, they are still completely dependent upon an external source of water for fertilization.

As in the mosses, a fern life cycle begins with the germination of a haploid spore. Mitotic cell divisions produce a flattened, heart-shaped disc, usually no larger than a thumbnail, called the gametophyte. On the ventral surface of a mature gametophyte are the structures that produce egg and sperm, the archegonia and antheridia, respectively. Usually, a single gametophyte bears both archegonia and antheridia. Also found on the ventral surface are rhizoids.

The gametes, egg and sperm, are produced by mitotic cell divisions. When a mature antheridium releases its sperm, the sperm swim to the adjacent archegonia, where they fertilize the eggs contained inside the archegonia. Through mitotic cell divisions, the newly developing sporophyte enlarges. The gametophyte usually dies off once the sporophyte reaches this stage. Although the fern sporophyte is initially dependent upon the gametophyte, it quickly becomes independent and the dominant generation of the fern life cycle.

As the sporophyte develops, it produces horizontal, underground stems called rhizomes. The roots and leaves of ferns are often finely divided, creating the appearance that a single frond is composed of many smaller "leaves." In many species, the undersides of the fronds contain the spore producing structures called **sporangia**. Clusters of sporangia are called **sori** (singular, sorus). A protective

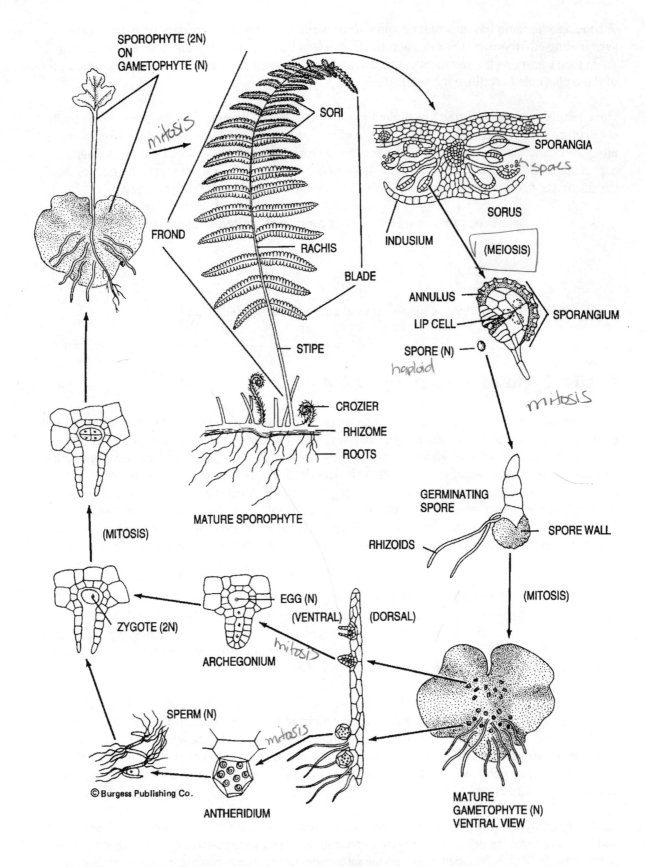

FIGURE 9-4. Life Cycle of a Fern

covering that grows over a sorus is called the **indusium**. It is inside the sporangium that spores are produced by the process of meiosis. In some species, a ring-like structure on the sporangium, called the **annulus**, opens the sporangium explosively, liberating its mature spores. The spores eventually fall to the ground where they will germinate into gametophytes, thus completing the life cycle.

1. What process is responsible for starting the sporophyte generation?

    mitosis

2. Name the structure in which meiosis occurs.

    sporangium

3. Name the structure in which fertilization occurs.

    archegonia

D. Life Cycle of a Flowering Plant, Figure 9-5

Two features that distinguish the flowering plants from other plants are the flower and fruit. It is thought that the flower has evolved primarily as a mechanism for luring animal pollinators, thus increasing the probability of fertilization. Fruits have evolved as structures to protect and facilitate the dispersal of young sporophytes.

In contrast to the mosses, where the gametophyte is the dominant generation and the sporophyte is dependent upon the gametophyte, flowering plants have reversed the roles.

The gametophyte of the flowering plant has been reduced to a barely multicellular, microscopic structure that is essentially totally dependent upon the sporophyte. In fact, the egg producing gametophyte is embedded in the tissues of the sporophyte, and has no "free living" phase. The only briefly independent phase is the sperm producing gametophyte, commonly called a **pollen grain**.

You will study the reproductive structure (**flower**) of flowering plants to help you understand the life cycle of flowering plants. The tissues that make up a flower are diploid, and therefore are products of the union of an egg and sperm. The plant body that supports and provides nutrients to the flower is also made up of diploid tissues. This diploid plant body is referred to as the sporophyte generation. As the name implies, its primary function is to produce spores. Did you know that most of the species of trees (except the conifers) are flowering plants?

In flowering plants, two kinds of spores are produced, those that result in the formation of male gametophytes, and those that form female gametophytes. The part of the flower that produces the spore that forms the female gametophyte is called the pistil. The pistil contains the following structures: **stigma, style, ovary, and ovule(s)**. Spores that produce male gametophytes are formed in structures called **stamens**. A stamen consists of an **anther** and a supporting stalk called the **filament**.

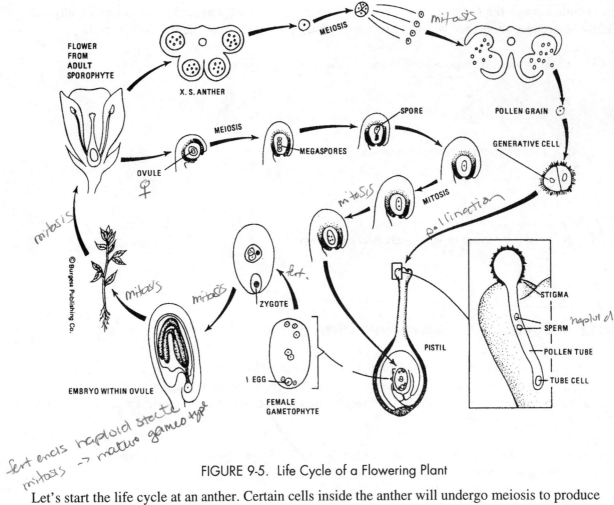

FIGURE 9-5. Life Cycle of a Flowering Plant

Let's start the life cycle at an anther. Certain cells inside the anther will undergo meiosis to produce haploid spores. In contrast to the mosses and ferns, the spores of flowering plants are not released from the sporophyte, but are retained in the structures that produced them. The spores undergo a limited number of mitotic cell divisions to produce structures called **pollen grains**. Pollen grains still contained inside anthers are actually immature male gametophytes because they have not produced a sperm nucleus. Eventually the pollen grains will be released from an anther and transported to the stigmas, the sticky terminal end of pistils. When a pollen grain attaches to a stigma, it is stimulated to undergo mitotic division, and to produce a structure called the **pollen tube**. The pollen tube grows down the style of the pistil until it reaches an opening to a female gametophyte. Contained in the cell that produces the pollen tube is the **sperm nucleus**. It is this haploid genetic material that will fertilize the egg of a female gametophyte.

Now let's consider the pistil. Inside a pistil is a structure called an ovule that contains a cell that will undergo meiotic cell division. One of the haploid cells produced during meiosis will eventually develop into a female gametophyte. The female gametophyte that is formed is only composed of a few cells, one of which is the egg **cell**. In contrast to the fern and moss gametophytes, the female gametophyte of a flowering plant does not have an independent phase. It lives out its entire life at the site of spore formation, the ovule.

When a female gametophyte has produced an egg cell, and a pollen tube has extended down to the opening of the female gametophyte, fertilization can take place by the union of the sperm and egg nuclei. With the formation of a diploid zygote, the gametophyte generation of the flowering plant ends. Subsequent mitotic cell divisions of the zygote produce an immature multicellular sporophyte called the embryo. The embryo grows for a period of time inside the ovule of the pistil, and eventually develops into a mature seed. A seed, as you know, germinates and grows into a structure we typically regard as the flowering plant, the diploid sporophyte. It is the sporophyte generation that produces those structures we call flowers. With the production of mature flowers the life cycle has come full circle.

1. In what structures does the process of meiosis occur?

   anther & ovule

2. What process marks the end of the gametophyte generation and the beginning of the sporophyte generation?

   fertilization

3. What process marks the beginning of the gametophyte generation and the end of the sporophyte generation?

   meiosis

4. Where do you find a mature female gametophyte in a flowering plant?

   ovule

II. GENERAL ORGANIZATION OF A FLOWERING PLANT, Figure 9-6

Flowering plants consist of five major organs: roots, stems, leaves, flowers, and fruits. The first three of these are considered vegetative organs, while the latter two are regarded as reproductive organs. Within the flowering plants, the type and number of these organs vary to a large degree. The garden bean plant, Figure 9-6, is a typical example of the arrangement of these structures.

Roots of plants generally serve as anchors to the substrate. Usually that substrate is soil, but sometimes it can be other plants. Many species, in the family Bromeliaceae (Spanish moss, wild pineapple, etc.) live on the surfaces of other plants. Plants of this type are called epiphytes. Root systems also serve as the primary means of procuring water and nutrients for the plant body. Some roots in certain species serve as food storage devices. Can you think of one species that uses its roots in this manner?

Stems are those structures that generally support the leaves, the primary photosynthetic machinery of the plant. The stems of plants can be greatly modified for functions other than support. For example, the Irish potato is a thickened, underground stem called a **tuber** that serves as a food storage structure. The onion is actually an underground stem that is surrounded by fleshy leaves. This type of structure is called a **bulb**.

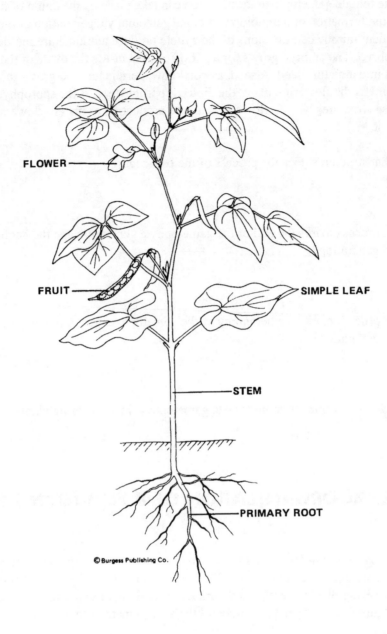

FIGURE 9-6. Garden Bean Plant

Leaves are extremely variable in their characteristics from species to species. Though their principal function is to act as photosynthetic factories, some are modified for other functions, such as the modified leaves of the pitcher plant (*Sarracenia sp.*) that are used to capture insects. In certain other species, some leaves are modified into protective spines.

Also supported by the stem are reproductive organs that are formed from highly modified leaves. These organs are called **flowers**. The **fruit** of a flowering plant forms from the part of the flower called the ovary. A fruit should not be confused with a seed, which is a mature ovule. Fruits contain one or more seeds.

III. ROOT STRUCTURE AND FUNCTION, Figures 9-7 & 9-8

Roots come in a variety of shapes and sizes, Figure 9-7. Some species have a well-developed main root that grows to great depths in the soil, while others have many shallow, spreading roots. An example of the former type would be the mesquite trees of the arid southwestern United States, while the many species of prairie grasses illustrate the latter type. Roots are the primary devices for obtaining water and nutrients from the soil. Therefore, it should not be surprising that the growth form of a plant's root system is highly correlated with the nature of water availability.

Root systems also play a major role in storing the excess amount of glucose produced during photosynthesis. Before storage, the glucose is polymerized into the storage product called starch. This excess food is often used as a reserve for winter, or more often as a large energy source to focus at a reproductive effort.

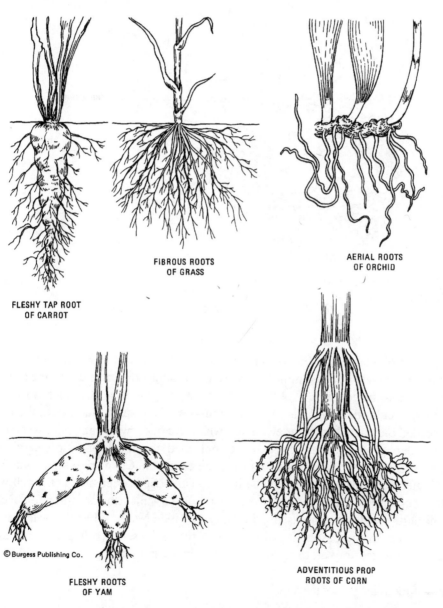

FIGURE 9-7. Root Types

Exercise 9 — Plant Biology

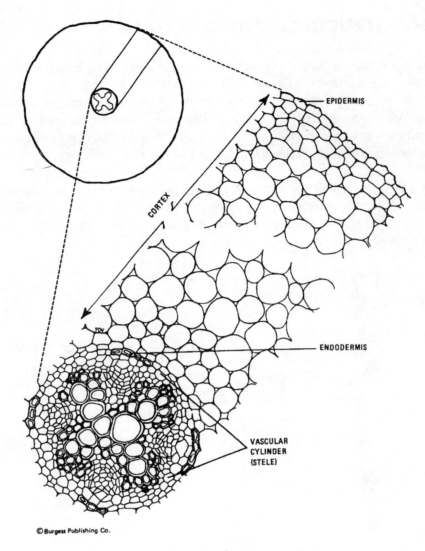

FIGURE 9-8. Cross Section of a Buttercup Root

Obtain a cross section of a buttercup (*Ranunculus sp.*) root. With the aid of Figure 9-8, locate the **epidermis** and the thick layer of cells immediately adjacent to the epidermis called the **cortex**. The purple-stained structures inside the cells of the cortex are starch granules. The inner most layer of the cortex is a singular, circular-band of cells called the **endodermis**. It surrounds a central core of vascular tissue, called the **vascular cylinder (stele)**, which conducts water and photosynthetic products. The endodermis plays a central role in the controlled movement of dissolved minerals into the plant. This is accomplished with the aid of a structure called the **Casparian strip**, which is a narrow band of waterproof material (suberin and lignin). This strip is deposited in the cell walls of the endodermis. Water entering the root through the epidermal cell walls can pass through the cell walls of the cells of the cortex without ever passing through a cell membrane. When the water reaches the endodermis, it is forced, by the Casparian strip, to pass through the selectively permeable endodermal cell membranes. Your instructor will explain this process in greater detail.

1. List three functions of root systems.

2. Explain how the Casparian strip in the endodermis of a root helps to regulate the movement of nutrients into a plant.

IV. STEM STRUCTURE AND FUNCTION, Figures 9-9 & 9-10

age of wood? twig

Obtain a woody twig from the counter, and with the aid of Figure 9-9 locate the structures in the following paragraph that appear in bold print.

Stems differ from the roots of plants in possessing leaves, and in having **nodes** and **internodes**. Nodes are locations where leaves emerge from the stem, and internodes are the areas, between two nodes, which lack leaves. **Flower buds** also form at nodes. Wherever a leaf emerges and subsequently detaches from a stem, a **leaf scar** is left behind. At the end of the stem is the **terminal bud**, which is covered by modified leaves called **bud scales**. The bud scales provide protection for **embryonic leaves** that lie beneath. The embryonic leaves are formed from a tissue called the **apical meristem**. Terminal buds represent the beginning of a year's growth. In spring, the terminal bud scales are shed, leaving behind bud scale scars at the site of their attachment. The apical meristem produces tissues that form an internode. At the end of a growing season another terminal bud forms. A single year's growth is the area between two adjacent bud scale scars. How old is your woody twig?

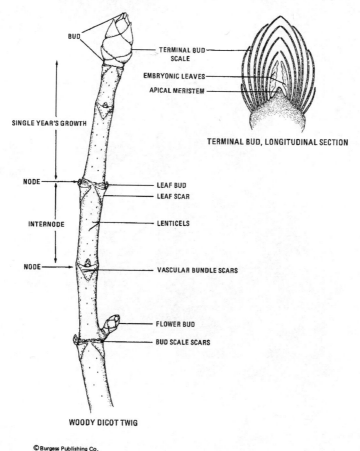

FIGURE 9-9. Woody Twig

Exercise 9 — Plant Biology 173

Yearly lateral growth (breadth) can easily be determined in older woody stems, such as in basswood (*Tilia sp.*) and oak (*Quercus sp.*). You have all probably heard of the annual growth rings of trees, but how many of you know how they arise and how to count them?

Obtain a prepared slide of a cross section of basswood, and observe it under your scanning power objective, Figure 9-10. The substance we call **wood** is water conducting tissue called **xylem**. The xylem is produced from a ring of mitotically dividing tissue called the **cambium**. Cells that arise on the side of the cambium closest to the center of the stem develop into xylem cells. During good growing conditions, the xylem cells develop into large cells, or what we call "spring wood." When Fall approaches or growing conditions are poor, the xylem cells that are produced by the cambium remain fairly small. We refer to this tissue as "summer wood." The small cells of the summer wood form a structure called the **annual ring**. By counting the number of annual rings you can determine the age of a tree. The oldest known tree is a specimen of bristlecone pine (*Pinus longaeva*) that is approximately 4,900 years old. Growth rings can also be used to shed light upon the growing conditions of the past. Your instructor will explain how this is done. Obtain a cross section of a wood block from the counter, and determine its age. You might find a stereomicroscope useful.

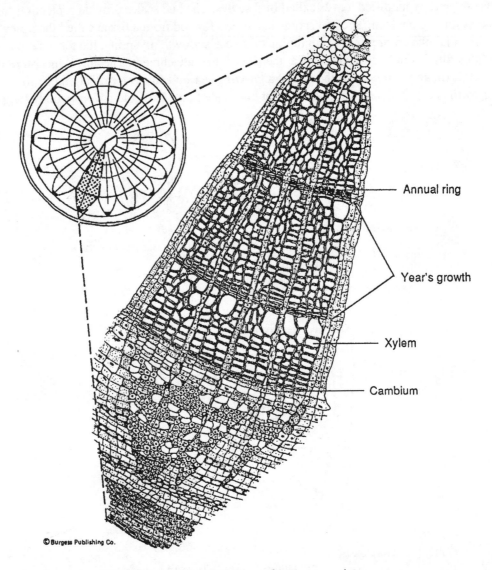

FIGURE 9-10. Cross Section of a Basswood Stem

1. What is a node? What is an internode?

   where leaves emerge from stems
   internode - area between nodes where there are no leaves

2. Describe how you determine one year's growth in a young woody twig.

   terminal buds

V. LEAF STRUCTURE AND FUNCTION, Figures 9-11, 9-12, & 9-13

Most flowering plants produce leaves from the nodes of their stems, although a few species, such as cacti, are leafless. Usually the form and arrangement of the leaves are species specific. A leaf may have three distinct parts: **blade, petiole, and stipules**, Figure 9-11. The blade of a leaf is the expanded area that contains the majority of the photosynthetic cells. The petiole is a stalk-like structure that supports the leaf on the stem, and the stipules are paired structures located at the base of the petiole. Not all of these structures are present in all plants.

Leaves can be arranged on a stem in one of three basic ways. If only one leaf occurs per node, the arrangement is referred to as **alternate**. If there are two leaves per node the arrangement is called **opposite**. The term **whorled** is applied when three or more leaves occur per node. What was the arrangement of leaves on your twig from the previous exercise?

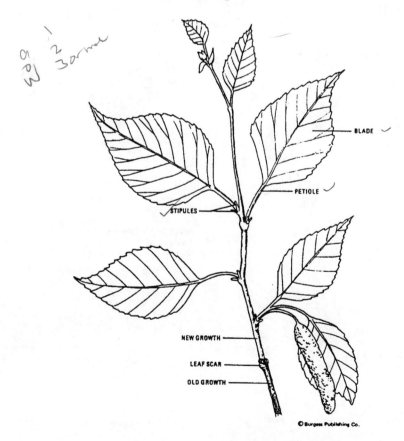

FIGURE 9-11. Leaf Structure

Leaves come in two types: **simple** or **compound**. Simple leaves have only one blade, whereas compound leaves have two or more blades called **leaflets**. Look at the plants on display in the lab and determine their leaf types.

Two characteristics of leaves that vary widely are leaf **margins** and **venation**, Figure 9-12. Leaf margins run from being smooth (e.g., monocot leaf) to being toothed (e.g., elm leaf). They may also be highly lobed as in mulberry. Venation, produced by the principal vascular tissue servicing the leaf, can be of three general types: **parallel, pinnate, or palmate**. As its name implies, parallel venation has tracts of vascular tissue running parallel to each other (e.g., monocot leaf). Pinnate venation has minor veins that arise on opposite sides of a main vein (rachis) of vascular tissue (e.g., elm leaf). The third type, palmate venation has three or more veins that arise from one point (e.g., simple dicot leaf). See if you can identify the types of venation of the plants on display.

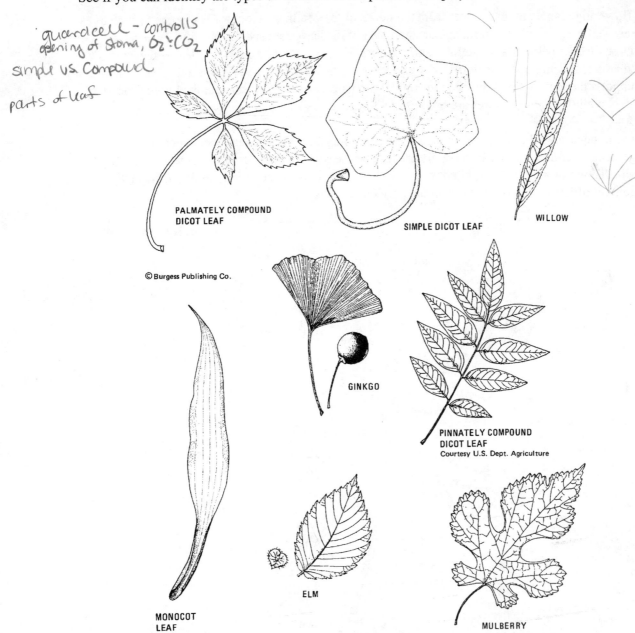

FIGURE 9-12. Leaf Types

Obtain a cross section of a privet (*Ligustrum sp.*) leaf, and observe it through your compound microscope. You should consult Figure 9-13 to help you identify the internal parts of the leaf.

Leaves are first and foremost photosynthetic factories. They generally have a large surface area over which the photosynthetic machinery is exposed to sunlight. Having a large surface area in a terrestrial environment has its problems because it increases the surface area from which water can be lost. Plants have structures that help to minimize water loss. One of these is the cuticle, which is composed of a waxy substance called cutin that covers the epidermis. Many of the plants that inhabit arid regions have well developed and thickened cuticles.

Notice from your cross section and from Figure 9-13, that just under the upper epidermis is a layer of tightly packed cells that are full of chloroplasts. This layer of cells called the **palisade layer** produces most of a plant's photosynthetic activity. Below the palisade layer is a loosely organized group of cells, called the **spongy layer**. The spongy layer is characterized by having large intercellular spaces that facilitate gas exchange. Also present in the leaf are bundles (vein) of vascular tissue (xylem and phloem). You should note that in the lower epidermal layer there are openings to the spongy layer called **stomata** (singular, stoma). These openings are flanked by two specialized epidermal cells called **guard cells**. Your instructor will explain the function of the guard cells. He or she will also discuss with you the seeming paradox of increased photosynthetic activity and the potential loss of large amounts of water in many of the grass species that inhabit hot, dry environments.

1. What is the function of the guard cells? Explain.

   Controls opening of stoma; O_2; CO_2

2. Name the three vein arrangements found in leaves.

   parallel, pinnate, palmate

3. Explain how it is possible for many plant species to carry out high levels of photosynthetic activity in hot, dry environments while their stomata are closed.

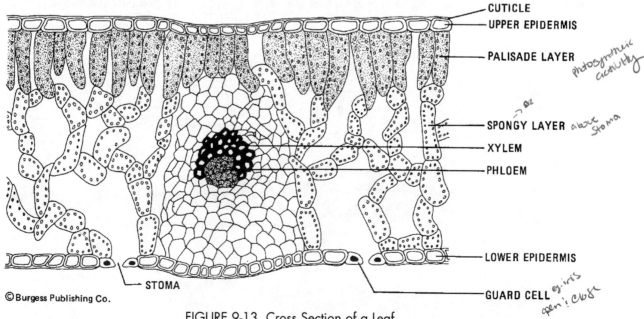

FIGURE 9-13. Cross Section of a Leaf

Exercise 9 — Plant Biology

VI. FLOWER STRUCTURE OF AN ANGIOSPERM, Figure 9-14

Flowers are made up of leaves that have become highly modified for reproduction. The flower, or what we might call the typical flower, has four basic parts: **sepals, petals, stamens**, and **pistil(s)**, Figure 9-14.

Sepals are the outermost part of the flower. They are generally green and leaflike, although they may be colored and modified. They generally are the outermost protective covering of the flower buds. Collectively the sepals are referred to as the **calyx**. The petals normally are found adjacent to the sepals, and collectively are called the **corolla**. The petals are often colorful and fragrant, serving to attract insects and other animals for pollination. Together the sepals and petals are called the **perianth**.

The portion of the flower that produces the male gametophyte (pollen grain) is called the **stamen**. A stamen is made up of the **anther**, a pollen producing structure that contains four separate sporangia, and a stalk supporting structure called the **filament**. Collectively the stamens are called the **androecium**.

The central part of the flower is occupied by the reproductive structures that give rise to the female gametophyte, the **pistils**. The **gynoecium** refers to the pistil or pistils collectively of a single flower. A pistil is composed of three parts: **stigma, style,** and **ovary**. The stigma is the pollen receptacle of the pistil. It is usually very sticky to help pollen adhere to the stigma. An elongated neck of the pistil, through which the pollen tube of a male gametophyte grows, is called the style. In some species the style is absent. Ovaries are structures that produce ovules. Ovules contain the cells that meiotically divide to produce spores. The spores then form female gametophytes, which in turn produce the eggs.

Flowers come in many shapes and sizes. Some are extremely elaborate structures, such as orchids, while others are much reduced and inconspicuous, such as flowers on elm trees. Even though there is a great diversity in floral structure, all flowers are modifications on a basic plan. For instance, some flowers have lost all sepals and petals, while others have modified sepals and petals into elaborate pollinating devices. Still others may have lost their sepals, but retain their petals. Some flowers have become so modified and reduced, that upon first glance, they appear to be just part of a single flower.

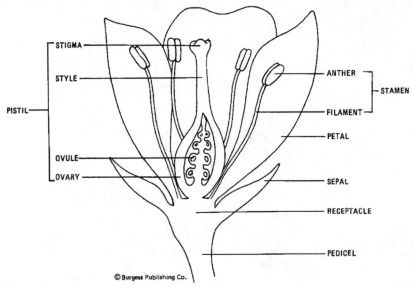

FIGURE 9-14. Basic Flower Structure

Upon closer examination we find what we thought was a single flower is really a dense collection of small, reduced flowers. This is the distinguishing feature of the group we call the composites (asters, dandelions, daisies, etc.).

The plants of the lily family (Liliaceae), which includes lilies, onions, tulips, yucca, etc., have flowers that typify the basic floral plan. That is, they have sepals, petals, pistils and stamens that are fairly unmodified and few. Each group of two students should obtain a lily from the back of the room, and the appropriate dissecting tools (razor and tweezers). You may want to use a dissecting microscope to view the details of the stamens and pistils. You should use Figure 9-14 as a guide to help you identify the parts of the lily flower. You will also find a fluid filled container containing an exploded dissection of lily and amaryllis flowers with the flower parts labeled.

In the space provided below, diagram and label the basic parts of a typical flower.

Flowers can be **perfect** (bearing both male and female parts) or they can be **unisex** (bearing male or female parts but not both). Plants that have male and female flowers on separate plants are called **dioecious**. Plants that bear both male and female flowers, but not perfect ones, are called **monoecious**, while plants that have perfect flowers are called **hermaphrodites**. Along the sides of the room you will find potted plants that have a variety of flower types. See if you can identify the flower types on these plants.

1. What is the function of a stamen?

   produces male gametophyte

2. In what structure does the female gametophyte form?

   ovule

3. Name one of the major functions of the petals.

   attract pollinators

The biological function of most flowers is the successful fertilization of its ovules with sperm cells from the pollen of the same species. Once this process of cross-fertilization has taken place, the flowers face the task of protecting the developing embryos from marauding herbivores.

Even though some plant species use wind or water as agents for the dispersal of their pollen, most flowering plants have recruited animals for this role. Most of us probably associate insects, such as the honey bee, *Apis mellifera*, with pollination. To be sure, the insects do play an enormous and important role, but many other animal species, such as birds, bats, and other mammals, are also essential reproductive intermediaries.

Central to understanding the pollination process is the concept of reward. Animal pollinators are not altruistically motivated out of the "goodness of their hearts" to move pollen for a plant from one flower to another. Rather, the animal must receive some type of reward, most commonly in the form of food, whether nectar, pollen, oil, or wax. Sometimes the reward may be a substance other than food, such as the aromatic substance collected from the *Anthurium* lily by male *Eulaema* bees. The perfume substance is used by the male bees during their elaborate courtship behavior.

Plants face an enormous task in trying to move their pollen from one flower to another of the same species. They must somehow ensure that their pollen is not prematurely dispersed, and that their pollen is not delivered to the "wrong" species. Most flowers must advertise that their pollen is ready for dispersal, and that the stigma is receptive. They do so by being conspicuous in one or a combination of color, shape, size or scent. Many plants advertise in such a manner that only one species will be attracted to its flowers. For example, out of the many species of euglossine bees only one species will be attracted by the scent of the *Gongora sp.* orchid. Another example is the orchid, *Ophrya speculum*, which mimics a female Scoliid wasp, *Gampsoscoha ciliata*, using a combination of color, scent, size, and shape. Males of the species make an attempt to copulate with these mimic flowers. In doing so, they pick up a load of pollen that can be delivered to another orchid during subsequent amorous visits. Other plants, such as species in the genus *Asclepias* (Milkweed family), are not as discriminating in their choice of pollinators and may use a host of animal species. All of these obstacles to cross-fertilization must be overcome by a plant while still maximizing the number of ovules that are successfully fertilized.

If time permits, your instructor will discuss the mechanisms of pollination in greater detail. In the lab, you will find examples of flowers that are specialized for different types of animal pollinators. See if you can determine the general type of animal that is used as a pollinator in each case. Your instructor will review the examples on display with you.

1. Why is a reward generally necessary for the pollinating species?

   > **HINT**: Think of the problem from an evolutionary perspective.

2. Do you think that there is one "best" pollination scheme among the flowering plants, or are the myriads of mechanisms equally "good?"

   > **HINT**: Again think of this problem from an evolutionary perspective.

VII. STRUCTURE AND FUNCTION OF FRUITS AND SEEDS

A. Fruits, Figures 9-15, 9-16, 9-17, & 9-18

Fruits are the mechanisms for protecting and dispersing immature plants, the seeds. They are structures that develop from the ovaries of flowers. A fruit is a matured ovary along with other flower parts that are regularly associated with it. Fruits may contain one or more seeds within the matured ovary. The wall of the ovary, the **pericarp,** consists of three layers: **exocarp, mesocarp, and endocarp**. The exocarp, the outer layer, often becomes the "skin" of the fruit (e.g., the red "skin" of an apple). Fleshy parts of fruit are generally derived from the mesocarp (middle layer), although other floral structures may also make up part of the fleshy material of a fruit. For example, the fleshy part of an apple is mainly made of the expanded base of the flower. Only the core of the apple is a true fruit, which is a ripened ovary. The inner layer, endocarp, can be modified in many ways.

Fruits are classified as **simple, aggregate, or multiple** depending upon the arrangement and origin of the **carpels** (modified leaves bearing the ovules inside the pistil). Simple fruits are those formed from one carpel or the fused carpels of a single flower. Some examples of single fruits are: plum, tomatoes, apple, and peas. A pea pod is an example of multiple carpels contained within a single pistil. Aggregate fruits are formed from multiple carpels contained in a single flower. Examples of aggregate fruits are raspberry, blackberry, and strawberry. The fleshy part of these aggregate fruits is formed from the expanded base of a single flower. Multiple fruits are formed from carpels from multiple flowers packed together into a tight cluster. As the ovaries from the separate flowers mature, they fuse to form a single fruiting structure. Pineapple and mulberry are examples of multiple fruits. Figure 9-15 illustrates how the structure of a flower from a particular species relates to the fruit that develops from the pollinated flower.

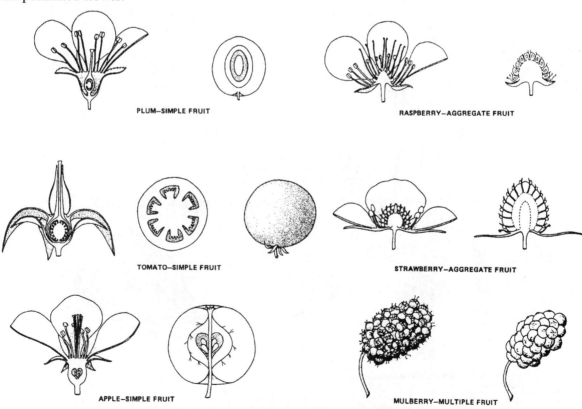

FIGURE 9-15. Flower/Fruit Relationships

Fruits may also be classified as soft and fleshy, or hard and dry. Apples, grapes, lemons, and watermelons are all examples of fleshy fruits. Dry fruits are classified into two groups: **dehiscent** and **indehiscent**. The ovary walls of dehiscent fruits break open at maturity to release seeds, Figure 9-16. The seeds of indehiscent fruits remain inside the fruits after the fruit has been released from the parent plant, Figure 9-17. You should look at the fruits that are on display, and dissect one or two that are available.

Fruits are structures that have become principally modified for the dispersal of seeds. The fruits of flowering plants show great diversity in their modifications for seed dispersal. Figure 9-18 illustrates just a few. Some fruits have become modified for wind dispersal, while others have become modified for dispersal by furbearing animals. Those of you who are dog owners are probably familiar with this reproductive strategy of plants. Many fruits have sweet, fleshy parts that attract species of highly mobile animals. For example, frugivorous (fruit eating) species of bats eat the fleshy parts of fruits, and then discard the seeds over wide areas.

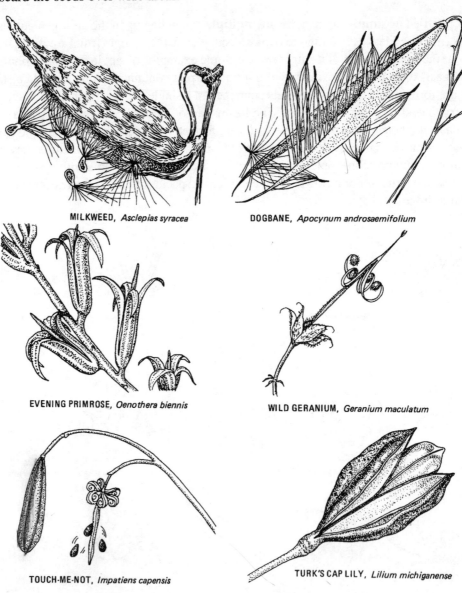

FIGURE 9-16. Examples of Dehiscent Fruits

182   Introduction to Biology: Laboratory Exercises

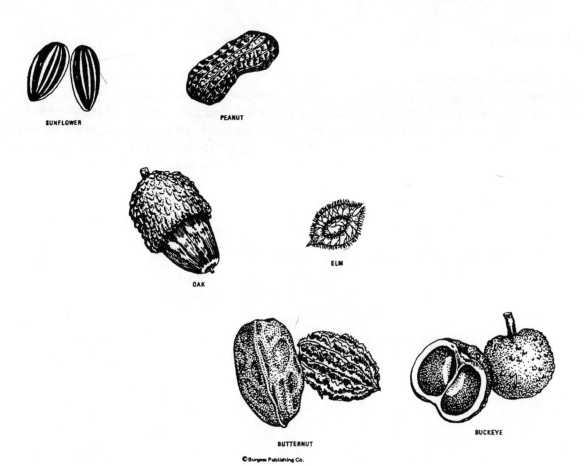

FIGURE 9-17. Examples of Indehiscent Fruit Types

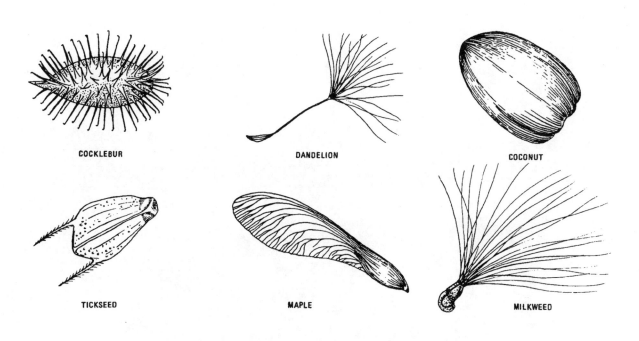

FIGURE 9-18. Fruits Illustrating Adaptations for Dispersal

B. Seeds

The **seeds** of flowering plants should not be confused with the reproductive structures called fruits. Remember that a fruit is a matured ovary that may contain one or more seeds. A seed, at a minimum, consists of an **embryo**, which develops from a fertilized egg, and a protective seed coat, which develops from the outer most layers of the ovule. Sometimes seeds contain, sandwiched between the embryo and **seed coat**, a nutrient layer of tissue called the **endosperm**. You may wish to dissect one or two of the seeds contained in some of the fruit examples you were asked to dissect.

1. Define the terms fruit and seed.

   f - matured ovary that may contain one or more seeds
   s - minimum consists of an embryo

2. Is the wall of an ovary sporophyte or gametophyte tissue?

3. Fruits appear to be adaptions for what purpose? Explain.

10 Plant and Animal Adaptations

INTRODUCTION

Why do birds that winter in southern latitudes migrate to temperate regions to breed? Why are some flowers colorful and elaborate while others appear uniformly drab and simple? Why are most mammals polygamous and most birds monogamous? Why do some species of birds lay one or two eggs while others lay considerably larger clutches? Why do some plant species produce hundreds of thousands of seeds annually while others produce only a few?

Evolutionary theory has provided us with insight into such questions about biological phenomena. Moreover, the theory is predictive, generates new insights, and stimulates further questioning. The theory distinguishes between two types of questions, those concerned with the immediate or proximate reasons for a biological occurrence, and those concerned with the ultimate or evolutionary reasons. The former deals primarily with mechanisms such as physiological or developmental processes. The latter are concerned with "why" biological phenomena exist in the first place. When trying to answer ultimate questions, it is necessary to ask, What is the significance of the phenomenon in terms of its contribution to or detraction from the reproductive success of an individual? The reproductive success of an individual is measured by the number of surviving offspring it leaves plus any contribution through its relatives.

For both plants and animals there exists a continuum of available reproductive opportunities. This simply means that an individual can produce zero offspring or as many as its reproductive potential allows. Somewhere along this continuum is a reproductive outcome that will produce the maximum number of offspring for a particular individual, given the conditions of the local environment. A successful reproductive outcome does not necessarily mean large absolute numbers of offspring, although it may. What it means is that a reproductive outcome is successful if it leaves the maximum number of reproductively active offspring relative to the rest of the individuals in the population. To this end, it may be necessary for an individual to reproduce hundreds of thousands of offspring just to ensure that a few will survive to reproduce. Producing large numbers of offspring may not be necessary if others in the population are producing fewer. In some cases of sexually reproducing species, two surviving individuals might suffice. What is significant about two surviving offspring in sexually reproducing species? It is to this ultimate end, reproductive success, that the form and function of adaptations are shaped by the process of natural selection (differential reproductive success).

The purpose of this lab is to illustrate how the major processes and central concepts discussed in lecture and lab are relevant to our understanding of the phenomena we perceive. Central to this interpretation is the idea that extant (living) plants and animals are the products of an evolutionary history shaped by the process of natural selection and chance. It is within this evolutionary framework that we can explain how and why questions about biological phenomena.

This exercise will consist of a trip to a natural history museum to view life-size exhibits of biological communities, or a field trip to local communities. A discussion of various plant and animal adaptations will center on three types of communities: the prairie, deciduous forest, and coniferous forest associations of Minnesota. The distribution of these three major floristic associations in Minnesota is illustrated in Figure 10-1.

OBJECTIVES

1. Be able to explain the difference between a proximate and an ultimate reason for a biological occurrence.
2. Be able to explain briefly the role of energy, nutrients, and fire in deciduous, coniferous, and prairie communities.
3. Be able to cite examples of plant adaptations to fire.
4. Define:
   climax community   reproductive success
   shade-tolerant   biological community
   shade-intolerant

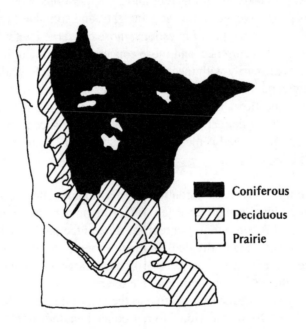

FIGURE 10-1. Vegetation Map of Minnesota

A **biological community** is any assemblage of plant and/or animal populations occupying a specific area. Beyond this simple definition are two opposing views about the nature of community structure. One point of view is that communities have a particular composition. That is, there is some fundamental unit of organization present in each type of community that continues to reappear in space and time. Communities are viewed as superorganisms with selection acting at the community level.

186   Introduction to Biology: Laboratory Exercises

A contrasting interpretation is that a community is composed of discrete populations of species that have their own specific ranges of environmental requirements, and that these populations are found along gradients that correspond to these requirements. In this view, only species are capable of evolving and the community is simply a human abstraction, rather than a real product of natural selection. Which interpretation, if either, is correct has yet to be resolved.

The physical environment (temperature, precipitation, fire, topography, wind velocity, light, and soil) is an influential factor in determining the plant composition of a community. The dominant (largest and most numerous) plant species influence the kinds of subordinate flora that inhabit an area, while the total flora can modify and buffer the effects from the physical environment. The kinds and numbers of plants determine to a large degree which animals live in a particular plant association. In turn, animals can affect community structure as well. An important factor often overlooked or dismissed is the role that "chance" plays in shaping community structure.

I. THE ROLE OF ENERGY

Just as a constant input of energy is ultimately necessary to maintain and perpetuate any individual, it should not be a surprise that light plays a significant role in forming the structure of a community. Because trees convert such a large portion of the solar radiation reaching an area into chemical energy, they exert an immense influence on subordinate plants and on animals that inhabit an area. This is one reason the trees inhabiting an area are used to categorize communities.

A. Deciduous Forest

One major floristic association found in Minnesota is dominated by two deciduous tree species, sugar maple and basswood. Because the two species exert enormous influence on subordinate plant species, and because they have a high probability of occurring together, the entire community is called a **maple-basswood** association. This forest type is considered by some to be the **climax forest** in Minnesota. A community of this type is persistent and can perpetuate itself under present conditions. Climax forests are thought to be the last stages of plant succession.

Sugar maple and basswood trees illustrate, with some specific adaptations, the significance of energy in influencing species composition. The two species are **shade-tolerant**, i.e., they can produce offspring capable of growing under the shade of parent trees. This means that the offspring can often out-compete species that are not shade-tolerant. Another adaptation that helps their offspring in the competition for light is the production of an extremely dense canopy layer that greatly reduces the amount of light that can reach the forest floor. Saplings of both species can persist in shade-tolerant conditions for a considerable amount of time, while the saplings of **shade-intolerant** species may persist for only a few years. It is not uncommon to find sugar maple saplings one or two inches in diameter that are forty to fifty years old. Our understanding of these adaptations have contributed significantly to the formation of the climax community concept.

There is usually a pronounced stratification (layering) of the vegetation in most forests. This stratification is probably a response to the amount of light that reaches the different levels between the top of the canopy and the forest floor. Do you think that there would be a positive correlation between the photosynthetic pigment composition of plants and the wavelengths of light that reach the various species? Do you think selection would have favored maximum efficiency in capturing and converting the available light energy? Explain.

What relationship do you think may exist between the number of strata and the variety of animals that occupy an area?

Many subordinate species of trees in a maple-basswood association are also extremely shade-tolerant (hackberry, slippery elm, and ironwood). Some of these species have special vegetative reproduction mechanisms that allow them to compete with the more shade-tolerant sugar maple. For example, basswood readily produces sprouts from roots and the base of the trunk. This may allow it to persist for considerable periods, though no seedlings of basswood become established. If a break in the canopy occurs, the basswood sprouts will be in a good position to out-compete nearby sugar maple seedlings.

How can you account for the occasional occurrence of shade-intolerant species, such as aspen and white birch? Hint: storms and other local catastrophes. Explain.

B. Coniferous Forest

The forests of northeastern Minnesota are characterized by an abundance of coniferous trees, including white pine, red pine, jack pine, white spruce, black spruce, balsam fir, white cedar, and tamarack. This does not mean that hardwood species are absent. On the contrary, they make up a significant portion of the northeastern forest. Forests in this region are found along an entire continuum of environmental factors, from extremely wet to extremely dry, and from granite surfaces to loam and clay soils.

Forests in the northeast are often found growing on a surface that has exposed granite and a minimum amount of parent material and soil. A common tree found in this area is the white spruce, a shade-tolerant species frequently found in association with balsam fir, which is also shade-tolerant. The two species are similar in life form, i.e., they are slender and cone-shaped. They often occur together in extremely high densities and produce a closed canopy similar in effect to the maple-basswood forest. Because of the nature of the growth form of these trees, the amount of stratification is often less than in a deciduous forest.

The deciduous, white birch is another tree commonly found with spruce and fir. It is a fairly shade-intolerant species that has rapidly growing seedlings, and seeds that are wind dispersed over long distances. It is a common invader of burned areas. Propose an explanation that would account for the occurrence of a single white birch in a stand composed predominantly of spruce and fir.

White cedar is another species commonly found with spruce and fir. It is also shade-tolerant but usually does not attain as great a height as fir or spruce. White cedar is extremely prone to tipping because of its shallow root system. Tipped trees continue to grow by sprouting new shoots from branches and the main trunk. Because of this method of vegetative reproduction, an individual tree, produced from seed germination, may persist indefinitely and successfully compete for available sunlight.

In more xeric (drier) habitats it is not uncommon to find areas covered by high densities of pines.

C. Prairie

A prairie is a large, open, essentially treeless expanse of land whose vegetation is dominated by herbaceous plants. Herbaceous plants are usually small and contain little woody tissue. The dominant herbs on a prairie are the grasses. Growing with the grasses is an assortment of forbs (any herb other than a grass) and woody shrubs.

A striking feature of a prairie is its lack of diverse stratification. Do you think the small amount of stratification present in a prairie has an impact on the number and kinds of animals that can exploit an area? Explain.

Because of the absence of trees, the sun and wind have immense impacts on the herbaceous vegetation. Hot temperatures and windy conditions mean high rates of **evapotransporation** (loss of water from the soil by evaporation plus the loss by transpiration from the vegetation). Plants can reduce water loss by closing their stomata, but in doing so they reduce the amount of CO_2 that can enter their leaves. A conflicting situation is produced. Carbon dioxide is essential for photosynthesis, and water is essential for metabolic processes and structural support, but only one may be available at a particular time in the quantity that is necessary.

How then have plants inhabiting hot, dry, and windy regions avoided this problem? Some plants have evolved adaptations that help to reduce the rate of transpiration without closing their stomata. Among these adaptations are curled leaves and sunken stomata. Both serve to protect the stomata from wind. Other plants have extremely thick cuticles that reduce water loss from the epidermal surface. An interesting adaptation, to dry and hot environments, is an alternate photosynthetic pathway that allows a plant to use low concentrations of CO_2. This altered metabolic pathway is characteristic of many prairie grasses. Therefore, though temperatures are elevated and water is limited, certain prairie plants can continue converting large amounts of solar radiation to chemical energy.

II. THE ROLE OF NUTRIENTS

Not only are plants and animals in competition for energy, but they are also competing for an additional limiting resource—the matter (atoms and molecules) that makes up the structure of their bodies. The importance of carbon, nitrogen, oxygen, and phosphorous is apparent from their ubiquitous occurrence in biochemically important molecules (nucleic acids, protein, carbohydrates, etc.), but other mineral elements are no less essential for the maintenance of life. Some of these mineral elements are essential in the production of bone (calcium), the proper conduction of nerve impulses (sodium and potassium), the structure of photosynthetic pigments (magnesium), and the proper functioning of certain enzymes (sulfur, zinc, iron, copper, magnesium, potassium, and chloride). The availability of these essential nutrients can be an important factor that influences the distribution of plants and animals.

A. Deciduous Forest

The maple-basswood association is a nutrient rich community. Because of certain adaptations, the association can form, modify, and maintain soil conditions. Sugar maple and basswood are deciduous trees, i.e. they shed their leaves annually. Unlike certain other tree species (oaks), they do not retrieve nutrients from their leaves before leaf fall. Therefore, leaves rich in potassium, calcium, magnesium, and other nutrients are added to the forest floor each year. This abundant supply of nutrients, and a source of energy, provides ideal conditions for a rich soil fauna. The rich variety of soil organisms, in turn, creates a soil environment that is conducive to retaining nutrients and moisture. An abundant supply of moisture and nutrients for developing sugar maple and basswood seedlings and saplings are another reason this association can be so persistent in an area.

B. Coniferous Forest

In comparison to deciduous forest soils and prairies, coniferous forest soils are poor in nutrients. Many soil building organisms present in deciduous forests and prairies are absent or rare in coniferous soils because of the nutrient poor conditions. One major factor contributing to this condition is the acidic water produced during the decomposition of needle leaves. Conifer needles are low in basic nutrient salts but are rich in organic and inorganic acids. Water containing these acids percolates down through the soil and dissolves many mineral nutrients present. It then carries these nutrients away from the top soil and deposits them at lower levels. In certain cases, the minerals are deposited where they are unavailable to plants. In areas where the parent material is primarily sand, the nutrients may be exported away from the area because of excellent internal drainage. Again this may make the nutrients unavailable to plants.

C. Prairie

Of the three major plant communities in Minnesota, excluding swamp and marsh lands, the prairie is the most nutrient-rich community. The soil building process is most rapid in this type of community. The black topsoil characteristic of certain types of prairie is extremely rich in nutrients and organic matter. The elevated surface temperatures on the prairie, the increased moisture capacities of the soil due to the large amount of organic matter, and the nutrient pumping of the prairie grasses contribute to a high rate of soil building. Together these factors produce ideal conditions for a rich soil fauna that accelerates the decomposition of the extensive fibrous root systems of the grasses.

III. THE ROLE OF FIRE

The three floristic associations in Minnesota have been greatly influenced by the presence or absence of fire. The periodicity of fire was probably greatly affected by the amount of precipitation received by an area, the chance occurrence of lightning strikes, and the direction and velocity of wind. A factor that could have contributed to the severity of fires in an area was the interval between the fires.

A. Deciduous Forest

The maple-basswood forests of Minnesota largely owe their existence to the absence or reduced frequency of fire in their stands. Most species in this association cannot tolerate ground fires even if the fires are not severe. The low frequency of fires in this type of forest is probably due to the high moisture content of the litter on the forest floor and the small amount of combustible litter present during the dry summer period.

About the only adaptation to fire that the flora of this association possesses is the vegetative sprouting from the bases of some mature trees and saplings. Mature sugar maple does not sprout from the base, but young saplings damaged by fire may produce many suckers (strongly growing shoots).

B. Coniferous Forest

The role of fire in the coniferous forests of the northeast is probably best reflected in the adaptations to fire of the three indigenous pines of Minnesota (white, red, and jack pine). All are shade-intolerant, to some extent, and germinate best on exposed mineral soil. The presences of such soils and the absence of shade are conditions often produced by forest fires. The growth rate of the three species is rapid, taking advantage of reduced competition for available sunlight after fires.

The pines may be ranked according to their degree of shade-tolerance, with white pine the most tolerant and jack pine the least. They may also be ranked according to their occurrence on dry, sandy soil, with jack pine inhabiting the driest sites and white pine the more mesic (moderate moisture) areas.

White and red pine are tall trees; heights of over 200 feet are possible. Both species lose many of their lower branches in a self-pruning process that produces tall straight trunks that have very little combustible material. They are also extremely resistant to ground fire damage because of their fire resistant barks. Because of these conditions, crown fires were infrequent in white and red pine stands. Therefore, fires at varying frequencies served to establish and maintain white and red pine stands by preparing the soil for pine seed germination and growth, and by eliminating potential competitors.

Jack pine possesses some particular fire adaptations that make it a conspicuously fire-adapted plant species. It is a small tree compared to white and red pine, and it retains many of its lower branches in a dried condition. The seeds of jack pine are contained in resin-impregnated cones and retained on the trees in a closed condition for many years. Excessive heat and dryness are necessary before the cones will open and liberate their seeds. Forest fires have historically provided the conditions needed for seed liberation. The release of seeds from their cones does not occur until approximately 24 hours after the cones have been heated and the ground has cooled. Seeds sprout and grow rapidly on the bare mineral soil produced by a fire. Within five to ten years the trees may be sexually mature.

Because jack pine inhabits xeric areas, its stands are subject to frequent burns. Often these burns lead to severe crown fines that destroy the mature trees. Because of the fire adapted cones, a new generation of jack pine may be "seeded-in" after a fire has destroyed a mature stand.

C. Prairie

Why are true prairies essentially devoid of trees? In xeric prairies the answer may be that adequate moisture to support the growth of trees is severely lacking. This may be true of prairies that cover large portions of the dry west and Midwest, but how about the area of Minnesota that runs diagonally from northwest to southeast across the state, where the prairie and forest meet? This prairie/forest border receives nearly the same annual precipitation. Clearly there should be enough moisture in this area to support tree growth.

Prairie fire is the mechanism that is responsible for the absence of trees on the prairie side of the prairie-forest border. The prairies of Minnesota are characterized by a high rate of evaporation and periodic droughts. These two conditions, with a large amount of organic matter, increase the probability of fire. In short, the climate and soil conditions of an area ultimately determine, for the most part, the frequency of fires in an area.

Most prairie plants are adapted to fire. Many possess root or stem systems that can sprout new shoots immediately after a burn. In the long term absence of fire, many prairie plants can vegetatively persist under the canopy of coniferous or deciduous trees, until the recurrence of fire again establishes them as the dominant plant species of an area.

IV. SPECIFIC ADAPTATIONS

Note to instructor:

You may wish to illustrate some following categories of plant and animal adaptations. Compare these categories of adaptations among the three major floristic associations whenever possible.

- A. flower color
- B. flower morphology
- C. flowering schedule
- D. flower olfaction
- E. pollen dispersal
- F. seed dispersal
- G. clutch size
- H. bird vocalization
- I. antipredator adaptations
    1. cryptic coloration
    2. disruptive coloration
    3. conspicuous or patterned coloration
    4. detecting predators (visual, olfactory, and auditory)
    5. repelling predators (mechanical and chemical)
- J. food gathering adaptations

11 Origin of Life

INTRODUCTION

The Russian biochemist A.I. Oparin, in a small monograph published in 1924 in Russian (Oparin, 1924), was the first to propose that life arose as a consequence of the physical and chemical evolution of our planet. Oparin's ideas remained little known to those outside the Soviet Union until he expanded his thoughts on the topic into a larger volume published in 1936. This work was eventually translated into English in 1938 (Oparin, 1938). Another eminent biochemist of the day, J.B.S. Haldane, also wrote about the origin of life from nonliving matter (Haldane, 1929), but not in as much detail as Oparin.

Both men's work on this topic received scant attention until the early 1950's. It should be noted that there were other ideas about the origin of life on earth that predated Oparin and Haldane's hypotheses. Many major figures of history such as Plato, Democritus, St. Augustine, St. Thomas Aquinas, Copernicus, Galileo, Harvey, Francis Bacon and Descartes presented philosophical and theological discussions of the subject (Farley, 1977). Even Charles Darwin commented on the subject, when in 1871 he wrote a friend exclaiming "It has often been said that all the conditions for the first production of a living organism are now present which could ever have been present. But if (and oh! what a big if!) we could conceive in some warm little pond, with all sorts of ammonia and phosphoric salts, light, heat, electricity, etc. present, that a protein compound was chemically formed ready to undergo still more complex changes, at the present day such matter would be instantly devoured or absorbed, which would not have been the case before living creatures were formed." Perhaps not so surprisingly, Darwin's statement captures the flavor of Oparin and Haldane's hypotheses.

Oparin and Haldane conceived of an earth, recently consolidated, that had an atmosphere devoid of molecular oxygen, but rich in reduced gases. This is essentially the opposite of the current chemical state of the atmosphere. It was apparent to them that if the primitive atmosphere were reducing and devoid of oxygen, and the present atmosphere oxidative and rich in oxygen, the atmosphere of the earth must have gradually changed from its ancient state. The significance of this seemingly trivial fact cannot be over emphasized, for the very transition from a reduced atmosphere to an oxidative state created the conditions that allowed life to emerge. In fact, this transition has left its imprint in the form of oxidation/reduction reactions as the principal mechanism on which the biochemical pathways of all life are organized.

Somewhere along this transition, from a reduced to an oxidized atmosphere, the physical and chemical conditions must have been ideal for the generation of the basic molecular building blocks of life, the polymerization and self-assembly of these molecules into macromolecules, and finally the association and assembly of these molecules into volumes of space that preserved their integrity at a particular point on the redox scale of carbon.

The ideas of Oparin and Haldane addressed this ancient question of the origin of life without invoking the involvement of some vital force, which up to that time had been the hallmark of most origin of life hypotheses. Rather, the two men drew upon their knowledge of the physical sciences to formulate their ideas. Oparin and Haldane laid the foundation upon which all subsequent investigators of the origin of life have made their respective contributions.

It was not until the early 1950's that experimental investigations were undertaken at the University of Chicago, by Stanley L. Miller and Harold C. Urey, to test the hypothesis that the basic building block molecules of life could have been "spontaneously" generated in the primitive atmosphere of earth. An apparatus that could simulate the reduced atmosphere of earth was assembled by Miller, then a graduate student of Urey. The apparatus was a closed system that could be evacuated of any oxidative atmosphere, and then loaded with water and the reduced gases thought to be present some 3.8 billion years ago. A boiling flask and condenser were used to make the contents of the system circulate, while a high-energy spark discharge was passed through the simulated atmosphere to provide a source of free energy. Any chemical products produced in the atmosphere of the apparatus would condense and collect in the boiling flask. The chemical compounds in the boiling flask would then undergo secondary reactions to produce more complex organic compounds. Periodically, samples could be drawn from the flask for analysis.

In the initial experiment (Miller, 1953), the gases used to produce the simulated primitive atmosphere were hydrogen, ammonia and methane; the only other substance present was water. Within a single day of running the apparatus, some amino acids, the building blocks of proteins, began to appear in the boiling flask. As the experiment progressed, more kinds of amino acids began to show up in the contents of the boiling flask. The primary products identified in the first Miller experiment were amino acids, aldehydes and hydrogen cyanide. Subsequent experiments, run with other types and combinations of gases, produced other classes of basic building block molecules, namely sugars and nucleotides. To date, all the major classes of organic compounds found in biological systems have been synthesized in experiments of this nature.

To appreciate the significance of Oparin and Haldane's hypothesis, that is, that life arose in a primitive atmosphere composed of reduced gases, one must understand the concept of **oxidation/reduction**, and what happens to the element carbon, when it is exposed to different oxidative and reductive environments. The oxidation of a substance involves the removal of an electron from its structure, while reduction is the addition of an electron to a substance. When electrons are removed from a substance, they must be immediately transferred to another substance, since they do not exist as free entities. Therefore, we say that when one substance is oxidized another must be simultaneously reduced, therefore the term **redox** to describe oxidation/reduction reactions. You cannot have one without the other.

The significance of all redox reactions lies in the energy transformations that take place during the reactions. For example, a simple redox reaction is used to lift into orbit the enormous mass of NASA's Space Shuttle. Liquid hydrogen and oxygen are used as the propellants for the main engines of the Shuttle. The reaction of hydrogen with oxygen releases a tremendous amount of energy. The reaction looks like this:

$$2H_2 + O_2 \longrightarrow 2H_2O + energy$$

On the surface it is not readily apparent that this reaction involves the transfer of electrons. If we look at how the reaction proceeds we discover that electrons are indeed transferred from hydrogen to oxygen. The reaction takes place in three steps:

1. $2H_2 \longrightarrow 4H^+ + 4e^-$

2. $O_2 + 4e^- \longrightarrow 2O^=$

3. $2O^= + 4H^+ \longrightarrow 2H_2O$

The first reaction splits hydrogen into four protons (hydrogen nuclei) and four electrons. In the second step the four electrons from hydrogen are transferred to the oxygen, thus forming two ions of a highly reactive form of oxygen. The oxygen ions then react with the four protons to form two molecules of water. Thus, the overall reaction can be described as the oxidation of hydrogen with the simultaneous reduction of oxygen to form water, and the concomitant release of energy. In other words, hydrogen is "burned" in the presence of oxygen to obtain energy.

You might be wondering what all of this inorganic chemistry has to do with the chemistry of organisms? As it turns out, the chemistry of life is largely the chemistry of oxidation/reduction reactions of carbon compounds. The significance of these redox reactions is the energy transformation that occurs during the process. It is the free energy liberated by redox reactions that is harnessed by organisms to build, maintain, and ultimately perpetuate themselves.

To better understand the redox processes of organisms, and how oxidation/reduction emerged as the predominant method that organisms use to obtain and handle energy, let us consider the different redox states that can be assumed by the element carbon. Because carbon is a tetravalent element, that is, because it has four electrons in its outermost energy level, the element can form a maximum of four single chemical bonds. For example, the chemical bonding requirements of carbon could be met by forming four single bonds with four atoms of hydrogen, thus forming a molecule of methane:

$$\begin{array}{c} H \\ | \\ H-C-H \\ | \\ H \end{array}$$

In this state, the carbon atom is said to be reduced because hydrogen has been added to it. The addition of hydrogen or the removal of hydrogen from carbon is equivalent to the addition or removal of electrons. Hydrogen, composed of only a single proton and electron acts as an electron in disguise. When a single proton and a single electron are in close proximity to each other, their respective charges disappear to the observer. Therefore, methane (CH_4) represents the most reduced state of carbon because no further electrons may be added to it, while carbon dioxide (CO_2) is the most oxidized form of carbon. These two molecules form the extremes of a scale of redox potential for carbon. Between these two extremes other forms of carbon compounds are thermodynamically possible.

Many of you are probably familiar with the process that converts methane, the main component of natural gas, to carbon dioxide. The process is used to cook our food and heat our homes, just to name a few of its uses. The reaction proceeds as follows:

$$CH_4 + 2O_2 \longrightarrow CO_2 + 2H_2O + \text{energy}$$

Notice that in this redox reaction, electrons are removed from methane, and then placed on oxygen. Therefore, we say that the methane has been oxidized while the oxygen has been reduced. Again, the important point to remember is that energy is released by the oxidation of a reduced substance.

Now that some basics have been covered, we can consider the significance and the implications of Oparin and Haldane's ideas. Oparin pointed out that the primitive atmosphere of the earth must have been much different from the present atmosphere. Evidence for this assertion comes from many disciplines, such as geology, astronomy and cosmology. He hypothesized that the atmosphere would have been composed primarily of reduced gases, such as hydrogen, methane and ammonia. The oxygen on the planet would mostly have been locked up in nonvolatile silicate minerals or water. Therefore, little or no free oxygen would have been present.

The significance of the absence of oxygen can be appreciated if we consider the redox scale of carbon, Figure 11-1. The diagram illustrates that in the most reduced primitive atmosphere (left side of scale) the only thermodynamically stable form of carbon would have been methane. That is, if any other forms of carbon were subjected to these conditions, the substances would have been unstable, and eventually would have been converted to methane. The right-hand portion of the diagram illustrates that in the present atmosphere the only stable form of carbon is carbon dioxide. In the contemporary atmosphere, any carbon compound, other than carbon dioxide, exposed to the oxidizing effects of oxygen will eventually be converted to carbon dioxide. Methane and carbon dioxide therefore represent the extremes of the redox scale of carbon.

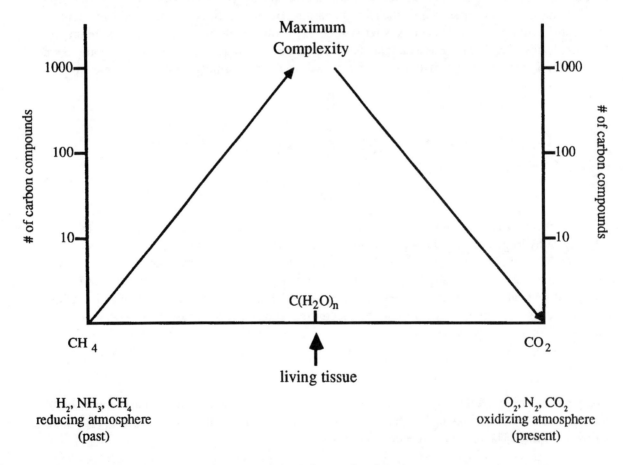

FIGURE 11-1. Redox Scale of Carbon

What other forms of carbon, if any, lie between these extremes? As you move from the extreme left to the right on the redox scale, notice the number of different forms of carbon that are thermodynamically stable increases to a point approximately at the center of the scale. Oparin realized the transition along the redox scale suggested a possible mechanism for the generation of the basic building block molecules of all life forms. He knew as the atmosphere of the earth lost the lighter gases, such as hydrogen, the atmosphere would have evolved to a more oxidized state. The biochemist hypothesized that if the gases in the atmosphere were subjected to sources of energy, such as U.V. light or lightning during this gradual transition of the chemical composition of the atmosphere, then chemical reactions would have occurred that could have generated many different types of carbon compounds. Since the atmosphere would have been devoid of strong oxidants, such as oxygen, the compounds would have been thermodynamically stable and could have accumulated in the water basins on the earth's surface. As J.B.S. Haldane stated in *Rationalist Annual* in 1929, without free oxygen to destroy these newly generated compounds, "the primitive oceans reached the consistency of hot dilute soup."

As the atmosphere continued to undergo its transition to a more oxidized state, a point would have been reached where the reactions that produced the "dilute hot soup" would no longer have been possible. Also, as the atmosphere became more oxidative, the compounds themselves would have become thermodynamically unstable. It is at this juncture Oparin said life must have originated. That is, some of the organic compounds must have associated and assembled in some collective manner in a restricted volume of space that offered protection from the oxidizing effects of the atmosphere. By assembling in a way that created an environment with a redox potential that was the same as their molecular structures, but different from the surrounding environment, these molecular assemblages could exploit the gradual transition to a more oxidized state.

At first glance, a move to a more oxidized atmosphere appears a dead end proposition because the molecular assemblages would literally be "burned" to simpler carbon compounds. Obviously this did not occur. Rather, these primitive molecular systems, confined in small volumes of space, must have been able to obtain adequate free energy from their surroundings to maintain their redox states, and thus their chemical and physical structure. By taking reduced compounds (food) from the surrounding environment, and by exposing these compounds to more oxidized molecules, they could control the rate at which these food substances "burned." In this way they could provide a constant supply of free energy to remain unchanged, i.e., maintain the reduced state of their structures. It was the thermodynamic considerations of the redox state of carbon that led Oparin to suggest that organisms are "living fossils," stuck at the midpoint on the redox scale of carbon, capitalizing on a 3.8 billion year old transitional event that allowed life to emerge, and which continues to allow life forms to build, maintain and perpetuate themselves.

The mechanism described above should sound familiar, for it is this very process that runs the industrial nations of the world. Reduced carbon compounds are exposed to the atmosphere (O_2) and ignited. The resultant oxidation of these reduced compounds to CO_2 and H_2O liberates fee energy used to build and maintain the structure of modern life. Just as we keep these types of fires under control by regulating the rate at which oxygen is exposed to the fuel source, our cells, and their enzymes, regulate the rate at which oxygen is allowed to interact with our food.

The experiments of Stanley Miller and scores of other experiments by subsequent investigators have all confirmed that the basic molecular units of life could have been generated in the primitive atmosphere of earth. Other individuals have demonstrated that many basic building block molecules can polymerize, and then "self-assemble" in the appropriate chemical and physical environment. Additionally, hypotheses on the origin of the genetic code, and its connection to protein synthesis also show promise in supporting the Oparin/Haldane hypothesis. Many questions remain to be answered, and still more asked, but a firm foundation has been established from which others may build.

During the next two lab periods, we will try to duplicate certain aspects of the Stanley Miller experiment that was first conducted in 1953. Because of the duration of the experiment (5 weeks) and the complexity of the apparatus, it was not possible for you to participate in the initial setup and run of the apparatus, but you will have the opportunity to collect and identify the contents of the collection vessel. Also, you will investigate another phenomenon that relates to the origin of life—chirality.

OBJECTIVES

1. Be able to explain how the transition from a reduced atmosphere to a more oxidized atmosphere played a central role in the origin of life.
2. Be able to outline the major stages in the origin of life from the formation of simple organic compounds to the formation of protocells.
3. Be able to explain how the redox scale of carbon is useful in understanding how life began.
4. Describe what is different between a mixture of amino acids from protein and the amino acids from a Miller apparatus sample.

MATERIALS

Miller apparatus (see preparation manual)
ninhydrin solution
bacterial culture plates
incubator
Proteus vulgaris culture
small vials
pulled Pasteur pipettes
amino acid standard solution
plastic bags
Bunsen burners
masking tape
grease pencils
insulated gloves
trivet

Miller sample
chromatography paper
D- and L-phenylalanine culture plates
inoculating loop
2% ferric chloride solution
hair dryers
chromatography developing jars
chromatography paper (Whatman No. 3)
molecular models
glass atomizer and tubing
small vials
hot plates
pencils and rulers
glass plates

I. THE MILLER EXPERIMENT

In 1953, Stanley Miller, then a graduate student of Harold Urey at the University of Chicago, assembled an apparatus that would attempt to test the Oparin/Haldane hypothesis that the basic building block molecules of life could be generated in a reduced gaseous atmosphere with natural free energy sources. The experiment was elegant in its theoretical simplicity and experimental design.

Basically, the experiment consisted of a simulated primitive atmosphere of reduced gases circulated in an enclosed glassware system while a high energy spark was passed between two electrodes. As chemical products were produced in the area of the spark discharge, they would condense and fall into the liquid water contained in the system. The concentration of the various products would continue to increase as the experiment progressed.

You are directed to Dr. Stanley Miller's original papers (Science, 1953 and Journal of the American Chemical Society, 1954) for a description of the apparatus and experimental procedures he used. Because the apparatus must run for approximately four to five weeks to obtain sufficient yields of amino acids, and because it is a complicated apparatus to assemble, you will only be concerned with the analysis of the contents of the collection vessel. But, you will have a chance to view the apparatus, and also extract from it the sample you will analyze.

The principal component of the Miller apparatus is a custom piece of glassware consisting of a boiling flask, condenser, and reacting chamber with electrodes, Figure 11-2. A gas manifold is attached to the main glassware via a vacuum stopcock. Each gas line of the manifold is attached to a specific gas tank (H_2, CH_4, or NH_3). Also, incorporated into the gas manifold is a vacuum line used to evacuate the atmosphere of the Miller apparatus, and all gas lines. A liquid nitrogen trap between the gas manifold and the vacuum pump is used to trap water vapor so it does not mix with the oil of the vacuum pump. An ammonia resistant pressure gauge is used to monitor the internal pressure of the apparatus while the experiment is running, and it is also used to load the gases in the particular ratio needed. The boiling flask is outfitted with a vacuum stopcock and spigot for extracting samples. A heating mantle is used to bring the system to a boil that causes the gaseous atmosphere to circulate. The spark between the electrodes is the source of free energy for the various reactions, and it is produced by a solid state induction coil. A water flow regulator is used to maintain a constant flow of water through the condenser.

The first step in loading the apparatus once the system has been assembled is to evacuate the atmosphere of the main apparatus plus all gas lines. When the system has been evacuated and checked for vacuum leaks, double distilled water may be added to the system. Next, the water must be degassed of oxygen by bringing the system to a boil. Again, the system must be vacuumed to remove any oxygen.

Once the system has been completely depleted of oxygen, the experimental atmosphere may be loaded into the apparatus. The pressure gauge may be used to monitor the amount of each gas added to the apparatus. Thus, any desired ratio of gases may be easily selected. When all gases have been loaded, the heating mantle and the water condenser may be turned on. A protective shield of acrylic should be mounted in front of the apparatus before any spark is introduced to the reaction chamber.

To analyze the contents of the apparatus, you will employ the method of paper chromatography. The procedure is based upon the fact that different organic molecules have different solubilities in various aqueous and nonaqueous solvents.

During the procedure, samples from the Miller apparatus and a solution of amino acid standards are absorbed into a piece of chromatography paper. One edge of the paper is inserted into a solvent that migrates through the sample deposit. Individual molecules begin to dissolve and move in the (non-aqueous) solvent. However, the cellulose fibers of the paper contain bound water molecules that form a stationary (nonmoving-) aqueous phase, while the solvent molecules form a mobile phase that is less soluble in the water phase.

Each type of organic molecule has a different solubility in the solvent and water. Molecules that are highly soluble in the solvent but poorly soluble in water will be preferentially separated (partitioned)

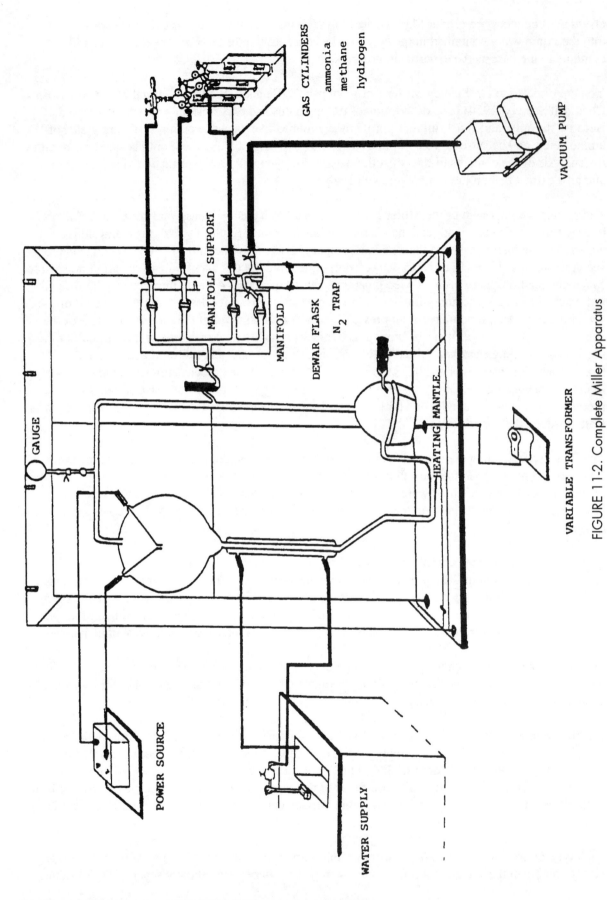

FIGURE 11-2. Complete Miller Apparatus

into the solvent phase and will move with the solvent. Molecules that have a higher solubility in water will preferentially partition into the stationary water phase and move at a slower rate. Each compound has a characteristic mobility (R_f) in a specific solvent system. R_f is the mobility of a compound relative to the migration of the solvent. That is, R_f is the ratio of the distance that a compound has migrated divided by the distance migrated by the solvent front. The observation that each compound has unique properties (R_fs in various solvent systems) has been applied to the purification and identification of many biological substances. If an unknown substance has the same R_f as a known (previously identified) compound in the same solvent system, then the compounds are probably identical.

You will prepare your chromatograms with a solution of "known" amino acids, and a sample from the Miller apparatus. When your chromatogram has been produced, you will spray it with the substance ninhydrin. Ninhydrin reacts with amino acids to give various colors of varying intensity for each type of amino acid (Hais and Macek, 1963). For example, the amino acid proline yields a yellow color when reacted with ninhydrin, while alanine gives a violet color.

After you have sprayed and developed your chromatogram with ninhydrin, you will calculate R_f values for the "known" amino acids, and then compare these R_f values to the R_f values that you calculate for the products from the Miller sample. You should use the comparative R_f value results, and also the similarities in color to identify the amino acids from the Miller sample.

Your instructor will also briefly discuss a column separation and atomic absorption technique used to identify the amino acids from the Miller sample. We have used this technique to obtain more accurate and quantified results at specific intervals during an experimental run. The results of such an analysis will be on display, and your instructor will discuss the technique and the results.

Procedure:

1. In a small vial, draw a sample (approx. 3-5 ml) from the Miller apparatus.

    BE CAREFUL. The sample will be hot, and it contains a high concentration of ammonia, which is caustic. Close the stopcock gently.

2. Cap the vial and take the sample to your laboratory room where you will prepare your chromatogram. On a piece of folded chromatography paper (3" X 11"), draw a thin pencil line about 1/2" from the bottom. Also, in pencil, label the side of your chromatogram with a name from your group and your laboratory section number. Place two small pencil dots on this line, one on each side of the fold. Exercise care in handling the chromatogram; touch only the edges of the paper. From your sample vial, extract some liquid with a pulled pasteur pipette. Spot this sample onto one of the dots. Blow dry the spot with a hair dryer until all water has evaporated. Repeat 10-15 more times, blow drying between each application. Label this side "sample." On the other pencil dot, using a different pulled pipette spot 3-5 drops of an amino acid standard solution (this solution should be labeled and at your table), and blow dry. Label this side of the paper "standard."

3. Once the sample and amino acid standard mixture have been applied to the paper and dried, use a pencil to label the paper at the top with the names in your group. Your instructor will collect the chromatograms from each group and place them in a large plastic bag. They will then be placed in a refrigerator until your next lab period, at which time you will complete the analysis. Because it takes over 5 hours to develop the chromatogram, the lab staff will start the chromatogram for you before you come to lab next period. The developing solvent is a 85% ethanol/water solution.

4. The first thing you should do at the beginning of the second lab period is to remove the chromatogram from the developing jar. The jars are located in metal wire baskets in the fume hood. Remove the chromatogram from the jar, and immediately mark the leading solvent edge with a pencil. You will need to know this position so you may calculate R_f values for the individual amino acids. To dry the chromatogram, hang it from the drying rack in the fume hood for approximately 15 minutes.

5. The next step in the preparation of your chromatogram is to spray it with the substance ninhydrin. Ninhydrin is a chemical that reacts with all free α-amino groups of amino acids to give blue/violet colored products. Each type of amino acid reacts with ninhydrin to give a characteristic color. These specific colors should aid you in the identification of the amino acids.

CAUTION: Ninhydrin is a toxic substance. You should avoid contact with your skin and inhaling any vapor. Wear protective gloves at all times when working with ninhydrin.

Your instructor will demonstrate how to spray your chromatogram. Be careful with the atomizer; it's fragile. All spraying should be done in the fume hood. Let the chromatogram air dry while hanging from clips provided on the drying rack in the fume hood. When it is dry to the touch, remove it from the clips. Place the chromatogram on a warm hot plate (2-3 setting) and set a pyrex dish on top of it. Note the order of spot development, and the spot colors. When a medium violet color is obtained with the standard, remove the chromatogram from the hot plate. Place the hot pyrex dish on an asbestos trivet when not in use. Use insulated gloves when handling hot glassware.

6. Calculate R_f values for the amino acid standards, and also for the spots produced from your Miller sample. Use the following formula:

$$R_f = \frac{\text{Distance an amino acid migrated}}{\text{Distance the solvent migrated}}$$

To calculate the distance each amino acid migrated measure the distance from the point of application, along a common vertical axis, to the center of each pigment spot. Each of these values (one for each spot) should be divided by the distance the solvent migrated, i.e., from the point of sample application to the leading edge of the migrated solvent. Listed below are R_f values for some common amino acids you might find in your Miller sample, and the colors produced when those amino acids react with ninhydrin. These values were obtained from running chromatograms on known amino acids using the technique just outlined in this procedure section.

Results from an amino acid standard sample:

Amino Acid	R_f value (EtOH)	Ninhydrin Product Color
Alanine	.28	violet
Aspartic Acid	.05	blue-violet
Glycine	.14	red-violet
Proline	.33	yellow
Valine	.47	violet

Results from the Miller sample:

Amino Acid	R$_f$ value (EtOH)	Ninhydrin Product Color
_____	_____	_____
_____	_____	_____
_____	_____	_____
_____	_____	_____
_____	_____	_____
_____	_____	_____

Questions:

1. What amino acids were you able to identify?

2. Were there any questionable results? If so, briefly explain.

3. Oxidation/reduction reactions are extremely important in explaining and understanding how life arose in a primitive atmosphere. Briefly explain how the chemistry of life came to be organized around oxidation/reduction reactions.

II. CHIRALITY AND THE ROLE OF ENZYME SPECIFICITY

If we examine the proteins of living organisms, we find that many of the amino acids that they contain are optically active. That is, when subjected to a plane of polarized light the plane of light is rotated in a specific direction. If we subject a mixture of amino acids extracted from the Miller apparatus to the same polarized light, no optical activity is observed. How can we account for this difference between the two mixtures of amino acids?

The answer lies in the spatial configuration of the molecules. Some substances, referred to as chiral structures, can exist in two forms whose structures are nonsuperimposable mirror images of each other. The word chiral is derived from the Greek word *cheir* that means hand. Therefore, we say molecules of this type display the property of handedness. When equimolar concentrations of each type are present no rotation occurs because one form rotates the light in the dextro (right-handed) direction and the other in the levoro (left-handed) direction, thus canceling the observable rotatory effect. If only one form, dextro or levoro, is present, then rotation occurs in only one direction, and the rotatory effect can then be observed. This phenomenon of optical activity is related to the stereochemistry of the compounds, that is, to the absolute configuration of the four different constituents in the tetrahedron around an asymmetric carbon atom.

All optically active organic substances may be related stereochemically to a single compound that has been arbitrarily selected to act as a standard for stereoisomers (substances whose molecules possess an identical structure but different arrangement of their atoms in space). Glyceraldehyde, the smallest 3-carbon sugar to have an asymmetrical carbon, is the standard reference molecule. The two possible forms of glyceraldehyde are, by convention, designated D and L. All amino acids that are capable of optical activity in living organisms, excluding some of those found as constituents of cell walls and antibiotics, are of the L configuration.

An apparent contradiction seems to be emerging from the previous statements. If the original mechanism that produced the basic building block molecules of life resulted in an equal mixture of stereochemical forms, then why do we find only the L form of amino acids in organisms? What caused or brought about the original selection of L forms of amino acids from the "primeval soup," and what process maintained the original choice?

The phenomenon of chirality produces some of the most intellectually stimulating and challenging questions and ideas about the origin of life. Some investigators have suggested that chance played the major role in the original choice of one stereoisomer form over its counterpart. Others have taken a point of view that is 180° from the idea of chance, and believe that the original selection was preordained by the physical and chemical environment in which the selection occurred. Whatever the reason, whether chance, deterministic chemistry, or something in between, it is clear to all sides of the argument that once the original choice was made, and there were proto-life forms that carried out chemistry linked to a genetic code, that natural selection was the process that amplified the choice and maintained it. The question of which point of view is correct is beyond the scope of this course, but we will look briefly at how enzymes, acting through the process of natural selection can maintain that original choice made some 3.8 billion years ago.

In the exercise today, you will investigate enzyme specificity as it relates to chirality, or in other words, the ability of an enzyme to specifically utilize the stereoisomeric forms of a compound. The reaction you will investigate is the bacterial deamination of the amino acid phenylalanine (an amino acid that is optically active) to phenylpyruvate. A deamination reaction simply removes ammonia from a substrate compound. The enzyme that brings about the reaction is L-amino acid oxidase. As its name implies, the enzyme works on amino acids that have the L configuration of the stereoisomers. The reaction is diagrammed in Figure 11-3 on the following page.

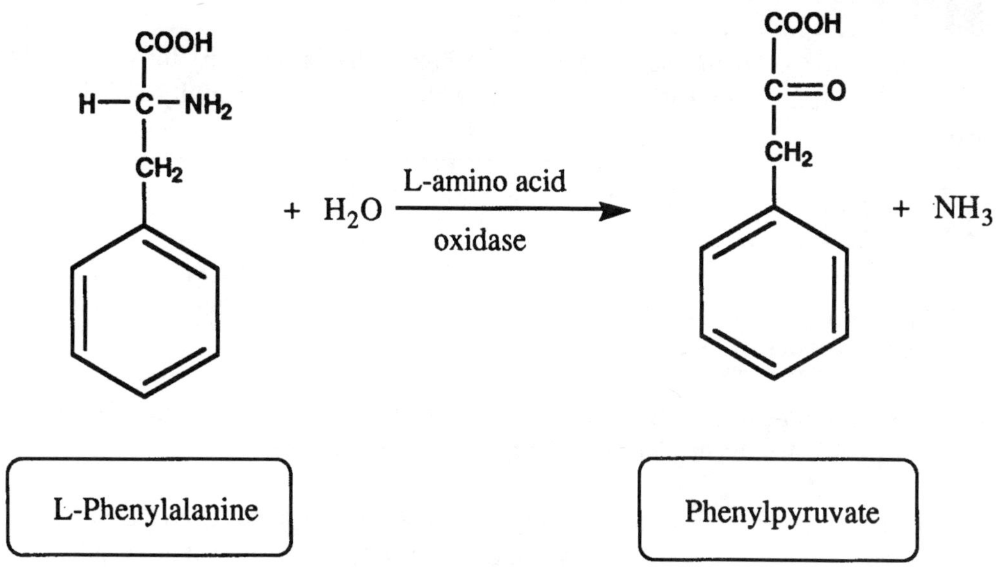

FIGURE 11-3. Deaminization of L-phenylalanine to Phenylpyruvate by the Enzyme L-amino Acid Oxidase

Rather than do this reaction in a test tube using chemical extracts made from organisms, we will allow the enzymes inside a living bacterium, *Proteus vulgaris*, to catalyze the reaction. To do this, it is necessary to grow the bacteria on a substrate that contains food. One of the foods that *Proteus vulgaris* uses is phenylalanine. This suggests that we should be able to prepare bacterial substrates that contain the different optical forms of phenylalanine, and then grow the bacterium on these plates to see if the organism can utilize (i.e., metabolize) the different stereoisomeric forms of phenylalanine. The test to check whether or not the reaction was carried out by the bacterial enzymes is a simple one. A few drops of a 2% solution of ferric chloride ($FeCl_3$) are placed on the bacterial plates after the bacteria have had a chance to grow. If deaminization has occurred, the drops of ferric chloride will turn green.

Procedure:

Note: Bacterial culture plates have been prepared for you in advance of your lab. The following culture plates differ as follows:

1. Complete media plus L-phenylalanine
2. Complete media plus D-phenylalanine
3. Complete media plus equal amounts of D- and L-phenylalanine

The types of plates are labeled with the numbers 1, 2 or 3. You will not be told which number corresponds to which type of substrate; that is for you to discover.

1. Each group of four students should obtain two culture plates of each type of substrate.

   > **CAUTION:** Do not remove the covers from the culture plates, doing so will contaminate the plates.

2. Your instructor will demonstrate the sterile technique that is necessary for you to follow while inoculating your plates. Failure to follow these directions carefully will surely botch the experiment.

   > **CAUTION:** Use care around the open flame.

3. After your group has inoculated your plates, stack them and bundle with tape. Label the tape clearly with your name, laboratory section, time inoculated, and date. Place your plates in the culture rack holders provided.

4. Your plates will be refrigerated for one day and then transferred to an incubator set at 37°C. The plates will then be incubated for 24 hours to promote growth.

5. When you return for your next laboratory class, collect your plates from the culture rack holders. Remove each cover from your plates, and add a few scattered drops of ferric chloride solution to the areas of growth you observe. After you have added the drops of ferric chloride watch for a color change. A green color indicates that deaminization has occurred. No color change means no reaction has taken place.

6. Record your results below:

Plate Type	Color Change + or −	Type of Substrate D-, L-, or DL
#1	_____	_____
#2	_____	_____
#3	_____	_____

References:

Oparin, A. I. 1924. *Proiskhozhdenie Zhizni (The Origin of Life)*. Moscow, Izd. Moskovskii Rabochii (in Russian).

Oparin, A. I. 1938. *The origin of life*. New York, Macmillan Company.

Haldane, J. B. S. 1929. The origin of life. *Rationalist Annual*, 148:3–10.

Farley, J. 1977. *The spontaneous generation controversy from Descartes to Oparin*. Baltimore, John Hopkins Univ. Press, 225 pp.

Miller, S. L. 1953. A production of amino acids under possible primitive earth conditions. *Science*, Vol. 117:528–529.

Miller, S. L. 1954. Production of some organic compounds under possible primitive earth conditions. *Journal of the American Chemical Society*, Vol. 77, No. 9:2351–2361.

Appendix I: Intraconversion in the Metric System

1 meter (m) = 100 cm
1 centimeter (cm) = 10 mm
1 millimeter (mm) = 1000 µm
1 micrometer (µm) = 1000 nm
1 nanometer (nm) = 10 Angstroms units (Å)

Conversion Charts for Metric Length Units

To convert from a unit in column A into any unit under column B, multiply by the power of 10 given in the appropriate column under B

A	B			
	nm	µm	mm	cm
Nanometer (nm)	1	10^{-3}	10^{-6}	10^{-7}
Micrometer (µm)	10^3	1	10^{-3}	10^{-4}
Millimeter (mm)	10^6	10^3	1	10^{-1}
Centimeter (cm)	10^7	10^4	10^1	1

Appendix II: Theory and Operation of the Spectrophotometer

Most of us can distinguish between a glass of water that contains one drop of blue food coloring and another that contains ten drops. Without knowing that one glass contained one drop and the other ten, we are able only to describe the difference in concentration in a relative manner, that is, "more" or "less." In other words, we are unable to describe the concentrations in a **quantitative** way. The reason that we can detect a difference in concentrations (in this case, the number of drops per glass) is that the two solutions transmit different amounts of light. They transmit different amounts of light because the "stronger" solution has more particles to absorb light energy than does the "weak" solution. Therefore, the absorbance and transmittance of a solution are a function of its concentration.

The Spec 20 can be used to measure quantitatively the amount of light absorbed by a solution at specific wavelengths. This information can then be used to determine the concentration of light-absorbing particles in the solution.

Figure App-1 is a schematic diagram of the light path through a Spec 20. Light is emitted from a tungsten lamp, passed through an entrance slit, and then dispersed into the spectrum by a diffraction grating. By rotating the diffraction grating, a specific wavelength of light can be made to pass through another slit. This opening has a small aperture that will allow only the selected wavelength to pass through it. The selected wavelength of light then passes through the sample solution. Light not absorbed by the sample passes through the sample and strikes a photosensitive tube that converts the radiant energy into electrical energy. The electrical energy can be measured by means of an electrical meter, and is expressed as a percentage of the total transmitted light.

The percentage of light transmitted by a solution is the difference in the electrical current produced by the solvent "blank" solution, and the electrical current produced by the sample solution. The % transmittance of a sample can be converted to absorbance by the following formula: $A = -\log T_f$, where T_f = **fractional transmission**, which is nothing more than transmittance converted to a decimal. Rather than using a formula to convert % transmittance to absorbance, an absorbance scale is provided on the Spec 20 meter. This allows rapid conversion of % transmittance to absorbance values. The main reason for converting to absorbance, rather than plotting % transmission, is because of the linear relationship of absorbance to concentration.

If a set of absorption values is generated from a series of solutions of known concentrations, and the results plotted, a curve ("**standard curve**") is produced that can be referred to in future experiments to determine concentrations of "unknowns" from absorbance values. Since the solute particles and the volumes of the solutions are the same, a valid comparison between absorbance values and concentrations can be made.

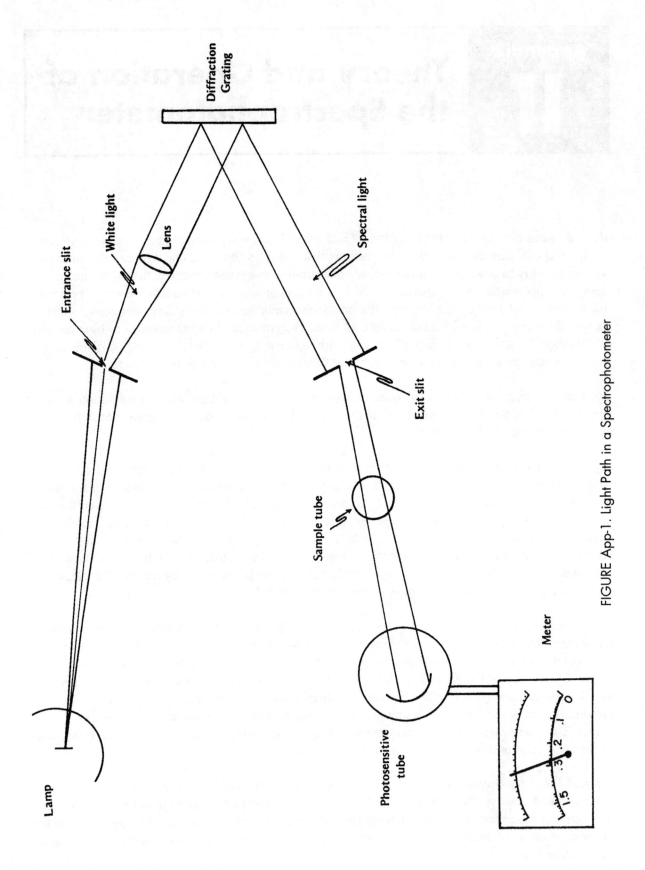

FIGURE App-1. Light Path in a Spectrophotometer

Because the molecules absorb different wavelengths of light preferentially, the Spec 20 can also be used to generate an absorption curve for a particular kind of molecule in solution. This information tells us which wavelengths are absorbed maximally, and which are absorbed minimally by a molecule when it is dissolved in a particular solvent. To produce an absorption curve, the concentration of the sample is held constant while the wavelength is varied. Absorption values are then recorded for each wavelength.

OPERATION, Figure APP-2

1. Turn on the Spec 20 by rotating the left knob in a clockwise direction. Allow the machine to warm up for a minimum of 15 minutes.

2. Before setting the scaled meter at 0% and 100% transmittance, select a specific wavelength by rotating the wavelength selection knob on the top of the instrument. The wavelength used will vary from experiment to experiment.

3. The left front knob on the instrument is used to set the meter reading to 0% transmittance- (0%T). This zero adjustment must be made without a cuvette (special glass sample tube) in the sample holder because the absence of a cuvette releases an occluder arm that prevents the light beam from reaching the photosensitive tube. The meter needle should always return to 0% T whenever the cuvette is removed. If it does not do so, reset the 0% T level with the left front knob.

4. To set the meter at 100% transmittance (100% T), a "blank" solution, one containing only the solvent, should be placed in the sample holder and the right knob rotated until 100% T is reached. Whenever a different wavelength is selected, the 0% and 100% transmittance levels must be reset. The 100% T level is always set with the "blank" cuvette.

5. The sample holder lid must be closed when taking readings.

6. Remove the "blank" after the 100% T level is set. Do not make any instrument setting changes after the "blank" has been removed.

7. The instrument is now ready to accept unknown samples. Insert a sample tube and read the value indicated on the absorbance scale (bottom scale). When the sample tube is removed, the meter should read 0% T. If it does not, reset the 0% T level. Insert the "blank" cuvette to check the 100% T level between successive readings.

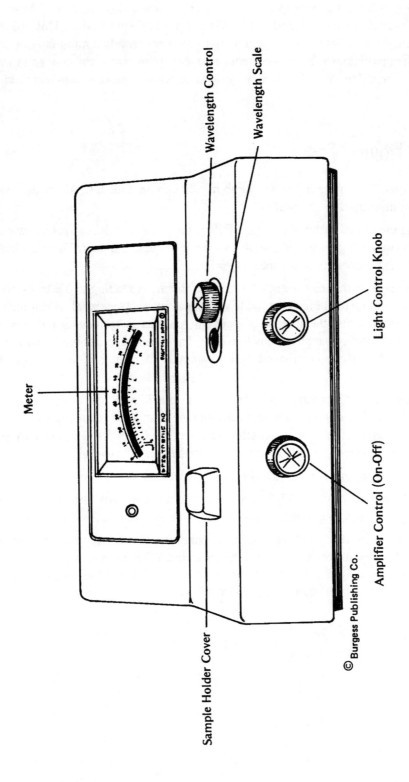

FIGURE App-2. Spectrophotometer (Spec 20)